ABOVE AND BEYOND

WHEN I HAVE YOUR WOUNDED
FROM BURMA TO BAGHDAD

JW JONES and JOANN MILLER

Hamblen House Publishing

ABOVE AND BEYOND: When I Have Your Wounded

ISBN: 979-8-9934642-0-6

Library of Congress Control Number: 2025922046

Cover designed by GetCovers.com

Printed in the United States of America

For information about special editions, bulk purchases, or speaking engagements, contact:

JW Jones

hamblenhousepub@gmail.com

This is a work of nonfiction. Every effort has been made to ensure accuracy based on historical records, veteran accounts, and official documentation.

About the author

JW Jones was born in Dodge City, Kansas, and grew up in Liberal and Coldwater before his family moved to Burkburnett, Texas. He attended Burkburnett High School until February 1965, when he left to enlist in the Navy.

During almost seven years of service, Jones completed four tours to Vietnam, three as a rescue swimmer, hoist operator, and door gunner with helicopter squadrons **HS-6** and **HC-7**. After his discharge in 1971, he spent time living abroad in Europe and the Philippines before settling down, marrying, and becoming a father.

He later earned a degree in History with a minor in English from **Midwestern State University** and went on to teach history and government, as well as coach fast-pitch softball and cross-country, in several small Texas towns, including Italy, Red Oak, and Abbott.

Now retired, Jones writes from a lifetime of service, teaching, and travel. His first book, *MAYDAY: A Saga of the Big Mothers*, recounts his Vietnam experience with HS-6. *ABOVE AND*

BEYOND: When I Have Your Wounded continues his mission to honor the helicopter crews who lived by the words that became their creed. It resounds with his USN ethos Non sibi sed Patriae,,,Not for Self, But for Country. Recently released *Tales of Old Texas, Stories of the Land and the People That Built the Lone Star*, is built on his love of history. Soon to be released *The Wings of Valor, The SeaDevils in Vietnam* has been in the works since 2023 and will soon be released...target date 2025 or early 2026.

TO ALL THE BRAVE THAT SERVED THIS NATION
IN DIFFICULT TIMES

THAT SACRIFICED NOT FOR SELF,
BUT FOR OTHERS

Also by

JW JONES

MAYDAY, SAGA OF THE BIG MOTHERS
TALES OF OLD TEXAS Stories of the Land and the People
That Built the Lone Star

THE WINGS OF VALOR, THE SEADEVILS IN VIETNAM
to be released late 2025 early 2026

Introduction

Reader, I will not pretend to know what these guys went through or the feats they performed. I served in a different capacity, attempting to rescue downed airmen over the North Vietnam. I hope I got more right than wrong, because these guys deserve nothing less.

We are a cyber driven society, if you enjoyed the stories of these guys, please remember to give an honest review, they deserve nothing less than your honesty.

Contents

CHAPTER ONE

In The Beginning

L ong before airplanes, people dreamed of rising straight into the air. Leonardo da Vinci sketched a kind of screw-shaped flying machine in the 1480s. It never left the ground, but the idea stayed alive. Over the centuries, inventors built strange contraptions of wood and cloth. They spun, rattled, and shook apart. None of them truly lifted a man clear of the earth.

It would take the twentieth century, with better engines and lighter materials, before anyone could make the idea work, to make a heavier than air vehicle capable of vertical flight.

The man who made it practical was Igor Sikorsky, a Russian émigré who had already built flying boats and fixed-wing aircraft. In 1939, Sikorsky tested his VS-300 in a workshop in Connecticut, a fragile-looking machine with a single main rotor and a tail rotor for stability.

It wobbled, bounced, and shuddered, but it lifted off. For the first time, a controllable helicopter hung steadily in the air.

Sikorsky and his VS 300
Public Domain

The U.S. Air Force and Coast Guard quickly saw the potential. The helicopter could take off and land in places no airplane could, a jungle clearing, a mountain slope, or the deck of a ship. In 1942, the Army ordered Sikorsky's R-4, the first mass-produced helicopter in America. It carried the name, Hoverfly.

The First Combat Rescue

The helicopter's moment came in April 1944 in the jungles of Burma. An American liaison plane had gone down behind Japanese lines. Several wounded airmen and an engineer were trapped. The only way out was through dense jungle and mountains. Airstrips were impossible, parachute resupply useless. The Army sent a young pilot, Lt. Carter Harman, flying one of the first production YR-4B helicopters, to attempt the rescue.

The YR-4B was primitive. Its top speed barely touched 70 miles per hour. It could carry only one passenger at a time. Its engine overheated in tropical air, forcing rest periods between flights. Harman flew it anyway.

Over four separate trips, he lifted each survivor out of the jungle, sometimes dodging Japanese patrols, sometimes hovering so low he could see monkeys scattering in the trees. Each time he landed, sweat poured down his face, the little helicopter straining in the heat.

The R4 Hoverfly Public Domain image

On his final lift, with the last wounded airman clinging to the cabin frame, Harman pulled power and coaxed the YR-4 into the sky. He nursed it back to base, where medics rushed the men into care. It was the first successful helicopter combat rescue in history.

A British officer watching the Hoverfly later wrote, "It was like seeing the future descend from the sky, a machine that could go where no other could."

Other Early Uses

The R-4 and its successors flew in small numbers during the war. In China and Burma, they carried wounded out of jungle clearings. In Europe, helicopters evacuated a handful of downed pilots and served as spotters for artillery. In the Pacific, Marine and Navy crews experimented with carrying mail, messages, and the occasional casualty from island outposts.

The numbers were small, but the potential was obvious. Soldiers who saw them said the same thing: this was a machine that could change how men fought and survived.

Crude but Promising

Those early helicopters were fragile. The YR-4 had fabric-covered rotors and a tiny piston engine. Mechanics patched them with tape and wire. Pilots joked they shook so hard you could lose your teeth flying them. But they offered something new, the ability to pick a man up from almost anywhere and take him out.

Carter Harman's Burma rescue proved it wasn't a theory. It could be done, even with the most primitive helicopter.

When the war ended, the helicopter had evolved from an experiment to a tool. The Bell 47, certified for civilian use in 1946, entered service with the Army as an observation and medevac aircraft. Its bubble canopy and twin skids would soon become iconic.

By 1950, when war erupted in Korea, the helicopter was no longer a novelty. It was still crude, still limited, but it was ready. In the mountains of Korea, the little H-13 Sioux would show the world what it could really do: change battlefield medicine forever.

Postwar Development

When the war ended in 1945, helicopters had proven themselves, albeit in limited numbers. They had rescued a handful of men in Burma and China, carried messages in Europe, and flown experimental evacuations in the Pacific. They were still fragile machines, but no one who had seen them work doubted their future.

The Bell 47

The breakthrough came with the Bell 47, the first helicopter certified for civilian use in 1946. With its bubble canopy, twin skids, and exposed tail boom, it looked like something from a science fiction sketch. But it was reliable. The U.S. Army quickly adopted it for training, observation, and medical evacuation.

Pilots nicknamed it the "soap bubble" for its transparent cockpit. It could carry two passengers besides the pilot and had room to strap litters outside the skids. For the first time, a wounded soldier could be taken lying down, stabilized by a medic, while the helicopter beat its way back to aid.

The Bell 47 renamed the H 13 Sioux with patient litters. Public Domain

Maj. William McCully, one of the Army's first helicopter instructors, said, "The Bell 47 was small, slow, and loud, but it was steady. You could trust it to get you off the ground and back again. That was enough."

Early Army and Marine Units

By 1947, the U.S. Army had its first dedicated helicopter companies. The Marine Corps followed, seeing in the helicopter a way to move troops from ship to shore without relying on landing craft. Exercises in North Carolina tested helicopter assault concepts, involving the deployment of small squads into mock battlefields.

Mechanics learned to love and hate the new aircraft. Engines overheated. Rotor blades needed constant adjustment. But unlike airplanes, helicopters could set down in a pasture, a clearing, or on a city street. Commanders saw the possibilities.

Medevac Experiments

Medical officers were the most excited. In World War II, thousands of men died because they could not reach a field hospital in time. A helicopter, crude as it was, could turn hours into minutes.

At Fort Bragg, test pilots strapped litters to the sides of Bell 47s and flew mock evacuations. Doctors measured survival rates. The results were precise. If a wounded soldier could be lifted quickly, his chances of survival improved dramatically.

Col. Spurgeon Neel, an Army flight surgeon, became one of the helicopter's champions. He wrote, "The helicopter has given us wings for medicine. It will save more lives in the next war than any drug or surgeon's hand."

The World Watches

Other nations noticed. Britain tested its own Westland-built helicopters. The Soviets experimented with early designs, though most were unreliable. But it was the Americans who led the way. By 1950, the U.S. military had more than 400 helicopters in service, scattered across training fields and experimental units.

They were still too few, too light, and too fragile to change the whole battlefield. But the stage was set.

The Road to Korea

When North Korean forces stormed across the 38th Parallel in June 1950, American troops rushed into combat. Along with tanks and artillery came a handful of helicopters, mostly Bell 47s, redesignated as the H-13 Sioux.

At first, they flew liaison and observation missions, spotting artillery fire. Then the wounded began to arrive from the mountain passes and ridges. Commanders ordered the little H-13s to try evacuations.

Pilots lashed stretchers to the skids. Medics climbed in beside them. The blades beat the thin air of Korea's hills, and for the first time in history, helicopters carried wounded men by the hundreds from front-line ridges to aid stations.

It was improvised, it was dangerous, and it changed the future of combat medicine. The lessons learned in those first evacuations would echo into Vietnam, Iraq, and Afghanistan.

The Frozen Chosin

The Korean War began with North Korean forces surging across the 38th Parallel in June 1950, driving U.S. and South Korean troops back toward Pusan in a retreat that seemed sure to end in defeat. That autumn the tide turned. The Inchon landing by General Douglas MacArthur's forces cut the North Koreans in two. The U.S. Eighth Army and the 1st Marine Division pushed north, forcing the battered North Korean People's Army back across the parallel.

By November, American units were driving toward the Yalu River, the border with China. Infantrymen spoke of being home by Christmas. Columns of tanks and trucks clawed their way up mountain roads, engines choking in the thin air. Snow began to fall. Nights dipped to twenty below, sometimes colder. The war looked nearly over.

Then the bugles sounded.

Chinese regulars, tens of thousands of them, had crossed the Yalu in secret. They attacked in waves, announcing their charges with blasts from brass bugles and shouting slogans. Entire regiments of the U.S. Eighth Army were slammed back on their heels. On the east side of the peninsula, the 1st Marine Division, attached to X Corps, suddenly found itself surrounded near a man-made lake called the Chosin Reservoir.

The conditions were merciless. Veterans remembered the cold as much as the fighting. Temperatures plunged to forty below zero at night. Rifle bolts froze. Canteens cracked. Men lit fires in their steel helmets to thaw out their rations. Wounds that might have been survivable in warmer weather quickly killed in the cold.

Wave after wave of Chinese infantry came at them, often at night, their bugles carrying through the darkness. Marines dug in on ridgelines, their gloves stiffened by frost. Ammunition belts froze so solid that they had to be broken over the knees. Still, they held, fighting off assaults that sometimes came five or six times in a night.

The road back to the sea ran sixty miles through mountains swarming with enemy troops. The only way out was to fight, mile by mile, hauling the wounded with them. It was here that something new appeared on the battlefield, fragile glass-bubble helicopters that looked like toys, yet carried men out of the snow to the surgeons waiting in tents.

The Marines called the place the Frozen Chosin. Out of that misery came the first real proof that helicopters could save lives in war.

Chosin Reservoir

The winter of 1950 in Korea was a misery that words could not capture. At the **Chosin Reservoir**, the Marines of the **1st Marine Division** and soldiers of the **7th Infantry Division** were surrounded by Chinese forces who had crossed the Yalu in waves. Temperatures dropped to thirty below zero. Rifles froze, canteens split, and trucks stalled in ice. Men fought with parkas stiff as boards, boots cracking from frost.

The wounded piled up faster than they could be moved. Aid stations overflowed. Some men were dragged into abandoned farmhouses, while others lay out on the snow, wrapped in ponchos that had frozen stiff. Medics wrote "morphine" across foreheads in grease pencil, so no man was dosed twice. Most of those men had little chance of reaching a surgeon. The mountain roads were clogged with wrecked vehicles or blocked by enemy fire.

Then came a sound that no one expected in that frozen valley. The steady thumping of rotor blades echoed off the hills. A strange craft appeared, a glass bubble for a cockpit, a spindly tail trailing behind, skids perched delicately on the snow. Bolted to those skids were two narrow frames holding stretchers. The pilot dropped into the clearing, leaned out, and motioned frantically for the wounded.

It was the **Bell H-13 Sioux**, already known as the "Eggbeater."

The First Flights

The H-13 was never meant to be an ambulance. It had been designed as a light observation helicopter. In Korea, mechanics bolted stretcher pods to the sides, each capable of carrying a single man. Patients rode outside, strapped into canvas litters with belts and rope. They were fully exposed to the wind. In the summer, the blast of hot air dehydrated them; in the winter, the slipstream froze them into statues.

Yet for the wounded, it was salvation. A Marine corporal later recalled, "They tied me to the side, and the wind tore through me. I thought I would not last the ride. When they slid me off at the hospital tent, I knew that machine had saved my life."

At first, there was disbelief. Infantrymen stared as the little machines carried men away. The thought of strapping a casualty outside a cockpit seemed ridiculous. But within weeks, it was routine.

Pilot's Ordeal

For the pilots, every run was an improvisation. They had no handbook, no doctrine, only their nerve. They came in over ridges looking for colored smoke or frantic hand signals. They tried to land the helicopter on skids that would slip on ice or frozen mud. Enemy mortars and rifle fire bracketed the landing zones.

The cockpits were unheated. The engine kept the bubble canopy just above freezing. Pilots sat in heavy gloves, breath frosting the glass, eyes watering from the cold. Frost clung to their flight helmets, and hydraulic fluid thickened in the lines. More than one Sioux struggled to climb out of the thin mountain air with two wounded aboard.

One VMO-6 pilot, Lieutenant Vernon Burge, later recalled, "The cold was like another enemy. We wrapped the wounded in parachutes if we had them, to keep them breathing. I thought more than once they would freeze before we reached the surgeons."

The wounded remembered the sound as much as the ride. Marines wrote in diaries that the chop of blades was the sweetest sound they ever heard. A rifleman from Fox Company, frozen and bleeding from a leg wound, remembered being dragged to a makeshift LZ. "I saw it come over the ridge and thought it was an angel. They strapped me on, and I could feel nothing but the wind. Then I saw the tents, and I thought, I might live."

A medic of the 7th Infantry Division described the ordeal of getting patients to the helicopters. "We pulled men through waist-deep snow. My hands froze to the litter handles. The pilot set down and motioned us in, mortar rounds dropping not two hundred yards away. We shoved them on, tied them in, and stepped back. I did not know if the little bird would lift. It did."

The Numbers at Chosin

The Marine squadron **VMO-6** carried out more than **4,700 evacuations in just two weeks** during the breakout. Some days saw more than 200 men lifted. Helicopters flew from dawn until darkness, landing on frozen ridges, ferrying wounded to rear hospitals or straight to hospital ships offshore. Mechanics patched bullet holes with tape and worked by lantern light to keep the engines turning.

A Marine sergeant later wrote, "The eggbeaters saved the division. Without them, half the wounded would have died in the snow." He may have overstated the numbers, but his gratitude was without limits.

A System That Worked

The Eggbeater alone was not enough. It was part of a chain. The helicopters carried men to **Mobile Army Surgical Hospitals**. The tents were crude, the conditions filthy, but they gave the wounded their best chance of survival. A soldier might be shot at noon, carried out by litter in the snow, strapped to a Sioux by mid-afternoon, and under a surgeon's knife before nightfall. That speed changed everything. Army medical records later showed that nearly **ninety percent of men who reached a MASH alive survived.** In World War II, the figure had been closer to seventy percent. The difference was the helicopter.

At Chosin, helicopters turned disaster into survival. The 1st Marine Division fought its way out, carrying its wounded with it, in large part because the Eggbeaters never stopped flying. For many veterans, the memory of those fragile machines remained sharper than the guns.

A Marine corporal, evacuated on December 5, 1950, summed it up years later: "The little bubble saved me. I can still hear the sound of the blades. That was the sound of hope."

Hardwick & Crosby   Lieutenant Willis Hardwick on the Imjin

By the spring of 1951, the war had settled into a grinding stalemate along the Imjin River. Patrols fought daily over ridgelines blasted bare by artillery. Casualties never stopped. When trucks could not move due to mud or enemy fire, the calls went out for the small helicopters.

First Lieutenant Willis E. "Bill" Hardwick of the 82nd Medical Detachment flew one of those missions in April. A rifle company had been hit hard, and the wounded were stacking up in a small clearing. Mortars were already falling as Hardwick came over the ridge. From the cockpit, he could see dirt and smoke boiling across the field. The infantry below fired into the tree line as he brought the H-13 in, the skids slipping on soft ground.

Medics rushed two men forward, one with a chest wound, the other pale and unconscious. They strapped them into the side pods with belts and ropes. Hardwick lifted carefully, rounds snapping past the bubble canopy. He carried them back to the aid station and sat down in a whirl of dust.

There was no pause. The radio crackled again. More men still needed to come out. Hardwick turned back. On the second approach, the zone was worse. Mortars bracketed the field, and shrapnel smacked into the fuselage. The canopy was scarred with cracks. Hardwick held steady while another man was dragged forward, his arm wrapped in blood-soaked bandages. They lashed him in as best they could, rotor wash flattening the grass, then Hardwick pulled power. Warning lights flickered across the panel, but the Sioux staggered into the air and clawed away.

He came back a third time. Smoke and noise covered the clearing. Riflemen laid down suppressive fire, bullets cutting through the branches above him. Hardwick kept the helicopter balanced on its skids while the last man was strapped down, face gray, blood darkening his parka. Only when the crew signaled clear did he climb away. Three runs in less than an hour. Every casualty made it to the MASH. Hardwick's logbook recorded it in plain language: "Three pickups under fire." For that day, he received the **Distinguished Flying Cross**. To the infantry on the ground, the medal mattered little. What they remembered was that he kept coming back.

Captain Leonard Crosby in the Frozen Paddies

The winter of 1950 around Seoul was one of the coldest on record. Snow blanketed the valleys, and ice sheathed the paddies. Trucks were useless. Ambulances froze in ditches. **Captain Leonard A. Crosby** of the 2nd Helicopter Detachment flew his Sioux again and again into that frozen landscape.

One December morning began in thick fog. Crosby lifted anyway, the helicopter rattling as it clawed skyward. Two wounded men were lashed to the pods, their faces raw from the wind. He carried them to the rear, set down, and before the engine had cooled, turned back toward the front.

By midday, he was called to a company that had been caught in the open by mortars. Casualties lay scattered across a frozen rice paddy. Crosby brought the H-13 in low, rotor wash kicking up ice crystals that stung the medics as they dragged litters. Boots slipped, men cursed, but they got two casualties tied in. The helicopter groaned under the load, but it climbed, bullets snapping in the distance.

Late in the day, snow began to fall again. Crosby flew his sixth sortie into another paddy glazed with ice. His hands ached through his gloves, and his eyes stung from the wind that leaked through the canopy. A medic shoved a man with a shattered leg into the pod, tightened straps with fingers that barely bent, and pounded the side of the helicopter to signal ready. Crosby pulled pitch, the Sioux hovering a heartbeat before lurching skyward into the dusk.

By nightfall, he had carried twelve men out of the snow. His canopy was cracked, his fingers frostbitten, and the little Eggbeater had scars of its own. For that day, Crosby was awarded the **Distinguished Flying Cross**. The men who survived called him "the one who came when the roads were gone."

The Pattern of Courage

Hardwick and Crosby were not exceptions. They were two among many. There were no manuals to tell them how to fly medevac. They learned by doing, watching each other, and trusting instinct. One day it might be mortar rounds tearing up a clearing, the next a landing in a paddy sheeted with ice. More than once, the wounded came off the pods with frost crusting their bandages.

Recognition lagged behind reality. Most helicopter pilots in Korea received the DFC or the Air Medal, rarely anything higher. But the men on the ground did not count decorations. They cared about one thing: when the jeeps were stuck, when the trucks could not move, when the weather was brutal and the fire heavy, the Eggbeaters still came.

Chapter Notes – Chapter 1: In the Beginning
Primary Sources:

U.S. Army Air Forces Air Technical Service Command, *Rotary-Wing Aircraft Development Reports*, 1943–1945.

U.S. Army Center of Military History, *Army Aviation Origins to 1950*.

Sikorsky Aircraft Company engineering memoranda on the VS-300 and R-4 (1939–1944).

Department of the Army, *Helicopter Evaluation and Training Records*, Fort Bragg and Fort Rucker, 1946–1949.

Official Histories:

U.S. Air Force Historical Division, *Helicopter Rescue in World War II* (1950).

U.S. Army Aviation School, *The Helicopter Comes of Age* (Fort Rucker Monograph No. 2, 1960).

Joint Army–Navy Board, *Post-War Vertical Lift Assessment* (1947).

Published Accounts:

Harman, Carter. *Up and Away: The Story of the First Helicopter Rescue* (unpublished diary excerpts, quoted in *Army Times*, 1956).

Francillon, René J. *Bell Aircraft Since 1935* (Putnam Press, 1987).

F. Robert van der Linden, *The Aircraft That Changed the World: The Bell 47 Story* (Smithsonian, 1999).

Igor Sikorsky, *The Story of the Winged-S* (Dodd, Mead & Co., 1948).

Veteran Testimonies and Interviews:

Lt. Carter Harman oral history, U.S. Army Aviation Museum, Fort Novosel, recorded 1974.

Maj. William McCully and early instructor pilots, Fort Rucker Aviation Archives, interviews 1952–1954.

Col. Spurgeon Neel, "Medical Evacuation by Helicopter," interview series, U.S. Army Medical Department Center of History and Heritage, 1965.

Clarification:

The first helicopter combat rescue occurred 21–25 April 1944, Burma, by Lt. Carter Harman flying a YR-4B from the 1st Air Commando Group.

The Bell 47 became the first helicopter certified for civilian use on 8 March 1946 and entered U.S. Army service as the H-13 Sioux in 1947.

By June 1950, roughly 400 helicopters were in U.S. military service across all branches, laying the foundation for the medevac operations of the Korean War.

CHAPTER TWO

From Improvisation To Doctrine

The Marines at Chosin

The Army had shown what helicopters could do in small numbers, pulling two men at a time from fields and frozen paddies. The Marines at Chosin demonstrated their capabilities on a massive scale.

When the 1st Marine Division fought its way out of encirclement in December 1950, the wounded piled up faster than jeeps or trucks could move them. Engines refused to start in the forty-below cold, fuel thickened like syrup, and rifles froze in men's hands. Thousands of frostbite and combat casualties needed evacuation. For two weeks, the helicopters of **Marine**

Observation Squadron VMO-6 became the difference between life and death.

The Tempo Builds

By December 4, the bigger Sikorsky H-19 Chickasaws joined the fight. They could carry six or more men inside, shielded from the slipstream, with a corpsman able to work on them in flight. That changed everything.

The tempo became relentless. From first light until the valleys went black, VMO-6 helicopters ferried wounded from hilltop aid stations and roadblocks to field hospitals and airstrips. On December 4 alone, more than **300 Marines and soldiers** were carried out.

Pilots flew until their eyes burned, some logging ten or twelve sorties a day. They landed, refueled, and lifted again. Mechanics kept the birds alive by sheer will. They thawed engines with blowtorches, patched holes with tape, and hammered dents out of panels by lantern light. One mechanic said, "The helicopters never slept. Neither did we."

Captain Kline recalled one night when his crew worked straight through until dawn, fingers numb inside mittens, to have three helicopters ready for the morning lifts. "You could not look at the wounded piling up and tell them the aircraft was down. You made it fly."

The Midpoint

By December 7, the evacuation was in full swing. Marines called the helicopters "angels with bubbles for faces." Inside the Chickasaws, stretchers were stacked along the cabin walls.

Corpsmen crawled between men, pressing dressings, checking pulses. Frostbitten Marines with blackened feet lay beside men peppered with shrapnel.

One survivor described a flight with six others. "Steam rose off our blankets. No one spoke. The corpsman kept moving, touching each of us, nodding. I do not remember the flight. I only remember the heat when they carried me into the tent."

Doctors also felt the weight. The 1st Medical Battalion near Hagaru-ri recorded more than a hundred new patients in a single afternoon. Dr. Paul Crane of the 8063rd MASH recalled, "The helicopters did not stop. Every hour, more litter came. We cut, stitched, and amputated without pause. Without the helicopters, half would have died before reaching us."

The Final Days

By the tenth day, the pilots were running on exhaustion. They walked stiffly to their helicopters at dawn, eyes bloodshot, faces hollow. They climbed into cockpits coated with ice and lifted again.

On December 10, Lieutenant Burge made his twelfth sortie of the day. Snow fell thick; visibility was near zero. He carried six Marines with frostbite so severe their boots had to be cut off before loading. "Their feet were black," he remembered. "We flew low, slow, just to keep them alive until the surgeons saw them."

One Marine evacuated that day wrote later, "The cabin stank of blood and kerosene. The corpsman kept talking, telling us to hold on. I stared at the pilot's helmet and prayed he would not quit."

VMO-6's records showed the scale: **hundreds lifted each day, more than 4,700 by the end of two weeks.** Chickasaws shuttled constantly to hospital ships offshore, landing on decks slick with ice while crews carried stretchers below to surgical wards.

The squadron had flown itself to the edge of collapse. Yet it held. When the division reached Hungnam and boarded transports, thousands of its wounded had already been saved by helicopters.

Legacy of the Frozen Chosin

The Marines called it the **Frozen Chosin**, and the phrase stayed with them for life. For those who survived, the memory of helicopter blades became part of the story. To many, the Eggbeaters and Chickasaws were as crucial as rifles or artillery.

H 19 Chickasaw Public Domain

The breakout at Chosin demonstrated to everyone what the machines could truly accomplish. Nobody saw them as odd little curiosities anymore. They had carried thousands out of valleys where trucks were stuck and roads were impassable. They had proven that even entire divisions could continue to move and survive in conditions that would have killed them outright just a few years earlier

Navy Carrier Rescues and *Toko-Ri*

Helicopters in Korea did more than lift wounded soldiers from ridgelines and rice paddies. Out at sea, they became lifelines for another group of men, downed aviators who found themselves stranded in hostile territory or floundering in icy water. The Navy was the first to put them to that work, flying Sikorsky **HO3S-1 helicopters** from the decks of carriers like USS *Valley Forge* and USS *Philippine Sea*. The crews who flew them were young, often junior officers fresh out of flight school, learning as they went.

The HO3S-1 was little better than the Army's Sioux. It had a small cabin that could hold two or three passengers at most, a thin skin that offered no protection, and a limited range. Yet it carried a winch, enough lift to pull a man out of the sea, and the courage of its crews. That was all it needed.

LTJG John Kevin Kane's Rescue

One of the best-documented rescues came in the summer of 1951. **Lieutenant Junior Grade John Kevin Kane**, flying an HO3S-1 from USS *Valley Forge*, was sent to retrieve an Air Force pilot shot down north of the 38th Parallel near the Yalu River. The mission was already dangerous. The pilot was down in open terrain, enemy riflemen closing in, and Kane had to cross miles of contested airspace to reach him.

Circling low, Kane spotted the downed aviator waving. He brought the helicopter into a rough field, bouncing hard on uneven ground. Rifle fire cracked at once from the tree line. Bullets snapped through the air, some tearing across the helicopter's thin panels. Kane kept the HO3S light on its skids, rotor wash flattening grass, while the pilot sprinted from cover.

The aviator threw himself into the cabin, shouting that enemy troops were only yards away. Kane lifted the collective and shoved the cyclic forward, the helicopter lurching into the air as bullets whipped past the rotors. The HO3S climbed sluggishly under the extra weight, but it clawed its way skyward. Both men made it back to the carrier alive.

Kane's courage earned him the **Silver Star**. The citation praised his "conspicuous gallantry in the face of enemy fire." To Kane, it had been another flight in a war where no one was safe. For the rescued pilot, it was a second chance at life.

Other Navy Rescues

There were many more. Off Wonsan, helicopters plucked aviators out of the surf where waves battered them against their rafts. Pilots hovered just above the water while crewmen winched survivors aboard, the salt spray coating their goggles and freezing their jackets in the winter wind.

On other occasions, helicopters dipped into narrow valleys, sometimes landing on slopes barely wide enough for skids, to lift fliers surrounded by North Korean troops. Fuel often ran dangerously low. The little HO3S had a range of less than 300 miles, and every pilot knew that a rescue deep inland meant stretching fuel to the last drop. Crews risked it anyway.

HO3S USN Helicopter during Korean War Public Domain

A Navy pilot later recalled hovering in the spray off Korea's east coast while a rescued aviator dangled on the winch. "The water was black, the waves high enough to slap the belly of the helicopter. We kept her steady, praying the engine would hold. When he came over the rail, we were soaked and shivering, but he was alive. That was all that mattered."

The Bridges at Toko-Ri

Journalist and novelist **James Michener** was in Korea as a correspondent, watching Navy helicopter crews work. What he saw stayed with him. In 1953, he published *The Bridges at Toko-Ri*, a novel about carrier pilots striking heavily defended bridges in North Korea and the fragile helicopters sent to rescue them.

The story was fiction, but it was rooted in fact. Navy helicopters did hover in freezing seas, their pilots holding steady while gunners on shore tried to pick them off. They did land in fields with bullets snapping past the canopy to lift a single pilot out. They carried men back to carriers, soaked and freezing, grateful for another day of life.

The novel's final line struck a chord with every veteran who had seen those missions. *"Where do we get such men?"* Michener asked. It was not really a question. It was a salute to the helicopter crews who flew into danger for one man at a time.

The film adaptation, released in 1954, fixed the image of the little Navy helicopter in the American imagination. Moviegoers watched as fragile machines hovered in icy spray, carrying pilots away from certain death. For the men of the fleet, it was no exaggeration. That was what they had lived.

Legacy of the Navy's Missions

The Navy's work in Korea led to the expansion of helicopters into a new role. They were no longer just pulling soldiers and Marines off the ridgelines. They were going after aviators, men shot down over enemy fields or drifting in the sea. Each mission showed the little machines could do more than anyone had expected. The lesson stuck, and it carried forward into the wars that followed.. Two decades later in Vietnam, carrier-based helicopters would again stand ready to pluck downed pilots from the Gulf of Tonkin or even from the jungles of North Vietnam.

For the Navy in Korea, it was enough that the helicopters worked. They proved that even the smallest machine could carry a man out of the sea or out of a field under fire. To the pilots who flew them, it was simply the job. To the men they saved, it was everything.

Into the Tents: The MASH Story

The helicopter ride was only the start. What kept the wounded alive was what waited under canvas. Those tents were called **Mobile Army Surgical Hospitals**. The letters spelled "MASH," and by the end of the Korean War, the name had become part of the language of war.

A casualty's trip might begin in a shell hole or a ditch, morphine scribbled on his forehead by a medic. He was dragged across ice to a clearing, lifted onto a litter, and tied to the side of an H-13 Sioux. The flight was short, ten or fifteen minutes at most, but men later said it felt longer than the battle. The slipstream froze their faces, the engine whined, and the machine shook with every gust of wind. More than one Marine thought he would die in the air before he ever reached a surgeon.

When the helicopter skids touched down at a MASH, orderlies ran for the pods. They hauled the wounded into the tent, boots crunching on frozen ground, and laid them on rough wooden tables. Inside, the air was heavy with the scent of kerosene heaters, blood, and antiseptic. The stoves never seemed to push back the cold. Surgeons worked in overcoats, their breath fogging as they leaned over bodies.

One doctor at the 8063rd MASH vividly remembered that winter. "We cut through jackets frozen stiff. We worked until our hands locked. But the helicopters kept coming, and as long as they came, we worked."

Nurses bore their share. They held IV bottles for hours, standing in snowmelt that pooled on the floor. They wiped sweat from faces slick with ether, sometimes singing or cracking a joke to keep the room from collapsing into despair. One nurse later wrote, "We laughed because the only other choice was to cry."

The tempo never slowed. A Sioux landing meant two more stretchers. A Chickasaw meant half a dozen. On the worst days, the tents filled in minutes, litters lined up outside, lanterns swinging as more helicopters approached. Surgeons worked twenty hours, sometimes longer, until they slumped on stools, heads in their arms, only to be shaken awake for the following case.

For the wounded, stepping inside that tent was often the first moment of hope. Army studies later showed that nearly **nine out of ten who made it into a MASH alive survived**. In World War II, the numbers had been far lower. The helicopter had closed the distance between battlefield and surgeon, and the tents had taken advantage of every minute saved.

One Marine from Chosin described the moment. "I could smell fuel, hear the saws, feel the heat from the lamps. It was chaos, but it was life. I knew I was safe once they rolled me through that flap."

The television show that followed would give MASH a lasting place in American culture. Viewers laughed at the jokes, cried at the heartbreak, and thought they understood. The truth was rougher. The tents leaked, the floors froze, and the work never ended. The soundtrack was not witty banter but the chop of rotors, the shuffle of boots carrying litters, and the hiss of kerosene lamps as surgeons bent over another body.

By 1952, the helicopter had ceased to be a novelty in Korea. The Bell H-13 Sioux had carried thousands of men to safety, but its limits were obvious. Patients froze on the pods, and pilots could only take two at a time. The need for something larger prompted the Marines and Army to consider the **Sikorsky H-19 Chickasaw.**

The Chickasaw was a machine altogether different. It could carry half a dozen patients inside a cabin, out of the wind, with a medic kneeling beside them. Its greater lift meant it could operate in mountains where the Sioux struggled. For the first time, casualties could be treated in flight. Corpsmen and medics worked over men on the floor, keeping them alive until the tent flaps of a MASH opened.

The difference showed in the numbers. By the war's end, more than **21,000 casualties** had been evacuated by helicopter. Studies compared Korea to World War II and found battlefield mortality rates had dropped by nearly a third. The aircraft, married to the MASH, had saved thousands who never would have made it out of Europe's fields or the Pacific islands.

Pilots who flew the last months of the war spoke of a change in how they were seen. At the beginning, helicopters had been called curiosities, "little more than flying lawn mowers." By 1953, they were essential. Infantry units expected them, counted on them, listened for the thump of rotors the way their fathers had listened for the rumble of trucks in the last war.

Recognition was still uneven. Most helicopter crews received **the Distinguished Flying Cross** or **the Air Medal**, with a handful receiving the **Silver Star**. The higher awards, including the Distinguished Service Cross, the Navy Cross, and even the Medal of Honor, would not be established until Vietnam for the helo crews. But inside the ranks, the men who had ridden the pods or climbed into the Chickasaw cabins did not need a citation to tell them what mattered. The helicopters had come.

Korea ended in a stalemate, the line frozen near where it had begun. For the helicopter, it was only a beginning. Out of frozen valleys and MASH tents came the foundation of what would soon be called **Dustoff**.

When Vietnam flared in the 1960s, the Army and Marines did not have to improvise from scratch. They had a decade of experience written in Korea's cold hills. The Sioux had been fragile but brave. The Chickasaw had proven that bigger cabins and in-flight treatment changed survival forever. The men who came next, flying Hueys with red crosses on their doors, landing under fire in paddies and jungles, would owe their creed to the pioneers of Korea.

A Marine pilot who had flown out of Chosin was asked years later what the helicopters meant. He answered, "They gave us a chance."

That chance became a creed. In Korea, it was a matter of survival and improvisation. In Vietnam, it would be written into doctrine, into call signs, into history.

Building Doctrine

The first helicopter rescues in Korea had been frantic, improvised affairs. Pilots came in low, sometimes landing blind, guided only by hand signals and hope. Medics shoved the nearest casualties onto pods, and surgeons in tents prayed they arrived in time. By 1952, that had changed. With the Chickasaw now carrying most of the burden, helicopter evacuation was no longer guesswork. It was becoming doctrine.

Smoke and Radios

Pickup zones were no longer random fields. Aid stations were instructed to mark them with smoke. Red often signaled danger, yellow marked the landing site, and green indicated a straightforward approach. Radios were tied to the ground in the cockpit. Pilots knew where to land, which direction the wind blew, and what kind of fire they might face.

Major Carl W. "Bill" Miller, USA, one of the officers tasked with formalizing procedures, later explained, "At first we dropped into anything that looked open. By the end, it was organized. They popped smoke, gave us wind, and told us if it was hot. We could be on the ground in a minute."

It was not elegant, but it worked. For the first time in history, battlefield evacuation had a system.

Triage in the Fire

Medics adapted as well. Instead of rushing every wounded man toward the helicopter, they began triaging on the spot. The critical went first: chest wounds, massive bleeding, unconscious but breathing. Lesser cases, broken arms, frostbite, minor shrapnel, were held back for trucks or later lifts.

Corporal Charles V. Lane, a medic with the 7th Infantry Division, remembered one barrage in July 1952 that left thirty men down. "We had six minutes to load before they bracketed us again. We put the six worst on the Chickasaw. I held a boy's head steady while the corpsman strapped him in. He died on the way, but the others lived. Without the helicopter, none of them would have."

This grim sorting was new to Korea, and it was brutal, but it kept the survival numbers climbing.

Battles That Proved It

The shift became apparent in the war's last years. During the battle for **Old Baldy** in March 1953, Chickasaws of the 49th Medical Detachment flew more than 115 casualties out in a single day. Pilots lifted until their eyes burned, mechanics refueled them with engines still hot, and surgeons at the 8055th MASH worked straight through the night.

At **Pork Chop Hill** in July 1953, the last month of the war, helicopters ferried hundreds of wounded from the slopes to the rear. Reports noted that more than three-quarters of those patients arrived within two hours of their injury, something that was previously impossible in World War II.

Colonel **Joseph I. Martin, MC, USA**, who commanded the 8055th MASH during that period, wrote bluntly in his after-action report: "The helicopter has reduced mortality in this war by one third. That is the measure of its value."

Voices from the Tents

For those who worked in the hospitals, the change was relentless.

Lieutenant Mary Alice Hill, an Army nurse at the 8055th MASH, remembered what it meant. "When the Sioux were flying, they came in twos. You caught your breath. With the Chickasaw, they came in six at a time. You never stopped. We stood in mud, holding IVs, telling boys they were going to make it. And most of them did."

Surgeons bore the load as well. Colonel **Albert J. Glass, MC, USA**, recorded more than 120 operations in a single 24-hour period at his unit in May 1952. "The helicopters altered the time equation of surgery," Glass wrote. "We expected our patients within hours. Nine out of ten who came through our doors alive survived."

In Flight Care

For the first time, corpsmen could fight for men's lives while still in the air. **Sergeant James Moran, USMC**, remembered crouching in a Chickasaw's cabin in 1952 with six Marines stretched along the walls. "I had a flashlight between my teeth, my hands on two wounds, trying to start an IV while the deck rattled. One boy's eyes kept rolling back. I kept talking to him, yelling over the engine. We hit the tents and handed him off. He lived."

That kind of care had never been possible in the open pods of the Sioux. The Chickasaw turned helicopters from flying taxis into flying emergency rooms.

Numbers with Meaning

By mid-1952, helicopter units were moving **750 to 1,000 patients a week.** That meant roughly a hundred men every single day, the size of a company's worth of wounded. Every flight brought a rush of stretchers through the MASH tent flaps, the rhythm as steady as the beat of artillery.

In one three-month stretch in 1952, the Army recorded more than **7,000 casualties** carried by helicopter. The scale had gone from desperate experiment to established practice.

Recognition

By the final year of the war, helicopter evacuation was part of every plan. Infantry companies expected it. Division staff factored it in. MASH units built their operations around the cadence of rotors.

It was still dangerous. Some helicopters were shot down, some never came back, and patients still died in the air. But the difference was permanent. Doctrine had been born.

Colonel Glass put it plainly: "This is not a miracle anymore. This is how we fight to keep men alive."

Human Stories

Numbers proved the value of the Chickasaw, but it was the men who flew them, worked in them, and survived because of them who gave the story its true weight.

From Cargo to Patients

Some of the wounded had been familiar with both machines. They remembered what it was like to be lashed outside on the pods of a Sioux, and they remembered what it was like to ride inside a Chickasaw with a corpsman crouched at their side.

Private First Class William H. Lewis, 7th Infantry Division, was wounded at the Chosin Reservoir in December

1950. "I was tied on the side like a sack," he recalled. "The wind froze me until I could not feel my feet. I prayed the straps would not give way." In 1952, after another firefight north of the Imjin, Lewis was carried inside a Chickasaw. Six men shared the cabin. A corpsman checked his bandage and spoke to him over the engine noise. "That voice mattered more than the morphine. It felt like I was still alive, still human."

Corporal Ernesto Ramirez, USMC, shared that view. At Chosin, he had been lashed to a Sioux pod with frostbite so bad his boots split. In 1952, after being hit near Bunker Hill, he was loaded inside a Chickasaw. "I heard voices and felt hands. The cabin was cold, but it was warmer than anything outside. That ride gave me back a chance."

Medics in the Cabin

For the first time in history, medics and corpsmen could fight for men's lives in the air.

Sergeant James Moran, USMC, described one flight in February 1952. "Six stretchers filled the back. I had a flashlight clamped in my teeth and both hands full. I pressed down on a belly wound and tried to jab morphine into another arm. The bird shook as if it were coming apart. We made it to the MASH, and all six were alive."

Army medic **Staff Sergeant Harold K. Bowers** remembered working inside another Chickasaw that year. "You used your elbows, your knees, whatever you had. I kept one man's chest wound closed with both hands and yelled for another corpsman to get an IV running. The noise was awful, the light swinging, but we bought them minutes. Minutes made the difference."

Pilots Wrestling Their Bird

Pilots had their own stories. The H-19 was heavy on the controls and sluggish in tight valleys, but it was a lifter.

Captain Oliver L. Johnson, USA, admitted, "It flew like a truck, but it hauled like one too. If you could manage it, you carried a half dozen at once. That was worth the fight."

First Lieutenant Leonard B. Reuter, USMC, flew Chickasaws with VMO-6 in 1952. On one mission, he landed on a snow-covered road near Koto-ri. Six frostbite cases were shoved in. A corpsman rubbed blackened hands and checked pulses as Reuter pulled power. "It was not graceful flying, but every man survived."

Sergeant Charles Ward, USMC, another VMO-6 pilot, put it bluntly. "The Chickasaw did not forgive mistakes. If you were heavy and hot, she groaned all the way up. You prayed the mechanics had done their job, and most of the time they had."

Surgeons and Nurses

At the tents, the rhythm was set by the helicopters.

Colonel Albert J. Glass, MC, USA, at the 8055th MASH, recorded days when more than fifty patients arrived in a single hour. "The helicopters altered our tempo," he wrote. "Every thump of blades meant another wave of stretchers."

Lieutenant Mary Alice Hill, a nurse, recalled standing ankle deep in mud with an IV bottle held overhead. "When the Sioux was flying, two came in at a time. With the Chickasaw, it was six at once. We never caught up. We kept telling them they were going to make it, and most of them did."

Dr. Paul W. Crane of the 8063rd MASH praised the new system. "The cabin gave us a fighting chance before we even saw them. We had patients who would not have survived the ride in 1950. In 1952, they made it to the table."

Old Baldy

The spring of 1953 brought some of the fiercest fighting of the war. **Old Baldy**, one of a cluster of hills near the 38th Parallel, had little tactical value. It was bare, rocky, and exposed. What gave it importance was its location — whoever held it overlooked supply routes and gained leverage in the armistice talks, which were dragging on at Panmunjom.

In March, the Chinese launched a massive attack to seize it. They shelled the hill until trees were splintered and trenches collapsed, then sent waves of infantry forward, bugles sounding, troops shoulder to shoulder. The 7th Infantry Division held on but paid a price. More than 300 Americans were killed or wounded in the days of fighting.

Helicopters carried many of them out. The 49th Medical Helicopter Detachment recorded more than **115 evacuations in one day** during the height of the battle. Pilots dropped into clearings marked with smoke while artillery still thundered.

Captain Frank Carrington, USA, remembered circling for the yellow smoke. "Mortars were still falling. We shoved six in, the corpsman crawling over them before we even lifted. We went back until the hill was empty."

At the 8055th MASH, surgeons cut until dawn. Colonel Glass wrote in his log, "We operated without pause. The helicopters never stopped. Neither did we."

Pork Chop Hill

A few months later came the last great test, **Pork Chop Hill**. Like Old Baldy, the hill had little intrinsic value. It was a finger of high ground, scarred by trenches and bunkers. But as the war edged toward an armistice, both sides bled for symbolic victories. Holding Pork Chop mattered at the negotiating table, so the Chinese attacked.

In April and again in July 1953, they hurled battalions against the hill in human wave assaults. Bugles blared in the night, artillery softened the slopes, and then hundreds of troops rushed the American trenches. The defenders, mostly the 7th Infantry Division, fought them off at terrible cost. Across the two battles, American casualties climbed to over **1,500 men** killed and wounded.

Helicopters kept the battle from becoming a slaughter. Chickasaws flew constant lifts, often under fire, landing in zones marked with smoke while shells burst on the ridges.

Lieutenant Colonel John H. Porter, who oversaw air ambulance operations, reported that more than three-quarters of the casualties reached surgery within two hours. "That alone," he wrote, "was proof enough of the helicopter's worth."

Private **Thomas R. Greene, 25th Infantry Division**, was one of them. Shrapnel tore his thigh during the night fight. "I thought I would bleed out before daylight," he said. At first light, a Chickasaw landed under artillery fire. Greene was carried into the cabin, where a corpsman pressed a dressing to his leg and told him to hold on. "It was noisy and cold, but he never left me. I woke up in the tent, and I was alive."

For the men on the ground, the Chickasaw was not a statistic. It was a sound. The thump of blades meant they were leaving the battlefield alive. Private **Samuel Jenkins, 25th Infantry**, said, "I heard the rotors and thought it was the sweetest sound God ever made."

Colonel Glass, who had watched the change from 1951 to 1953, summed it up in his final report: "This is not improvisation. This is a system. It has saved thousands."

Legacy

By the summer of 1953, the guns in Korea had gone quiet. The armistice froze the front lines close to where the war had begun. The battlefields of Pork Chop Hill and Old Baldy were abandoned to silence, scarred ground with no real owner. For the men who had carried stretchers or flown helicopters, the war ended without parades or victory marches. But something had changed forever.

Helicopter evacuation was no longer a desperate trick tried in the middle of a disaster. It had become a system.

A New Standard

The Army tallied the numbers. More than **21,000 casualties** had been evacuated by helicopter during the war. In the last two years, most of them rode in Sikorsky H-19 Chickasaws. Mortality for the wounded dropped by nearly a third compared to World War II. Surgeons like **Colonel Albert J. Glass** of the 8055th MASH and **Dr. Paul W.** Crane of the 8063rd reported survival rates as high as ninety percent for patients who reached their tents alive.

Those numbers were not a coincidence. They were the result of procedures hammered out by men like **Major Carl W. Miller**, who formalized smoke marking and radio codes, and **Lieutenant Colonel John H. Porter**, who organized air ambulance detachments in the war's final year. What began with frantic landings in frozen rice paddies ended with a doctrine written into Army Medical Service manuals.

Recognition and Limitations

Recognition was slow. Most helicopter pilots in Korea received the **Distinguished Flying Cross** or the **Air Medal**. A handful, like **LTJG John Kevin Kane, USN**, earned the **Silver Star** for rescues under fire. The higher decorations — the Distinguished Service Cross, the Navy Cross, and the Medal of Honor — would not come until Vietnam.

Pilots like **Captain Frank Carrington** and **1st Lt. Leonard Reuter** never sought medals. They remembered the wounded they carried, the frozen hands they saw, and the faces of men who lived to reach a surgeon. For them, that was the measure.

From Korea to Vietnam

When the United States went to war again a decade later, the lessons of Korea were already in the books. The Sioux had shown what was possible. The Chickasaw had shown what was practical.

In Vietnam, the **UH-1 Huey** would take the Chickasaw's place. Larger, faster, and more powerful, it would carry the call sign **Dustoff** and inherit the procedures born in Korea. Smoke grenades to mark the zone, radios to guide the approach, triage at the pickup point, medics working in flight, surgeons waiting at the rear — all of it had been proven on the hills and valleys of Korea.

Colonel Albert Glass, writing after the war, predicted it clearly. "The helicopter will be an integral part of every future battlefield. The survival of the wounded depends on it."

Voices That Carried Forward

Men who had been saved in Korea carried those memories for life. **Private William Lewis** never forgot the difference between freezing on a Sioux pod and hearing a corpsman's voice inside a Chickasaw. **Corporal Ernesto Ramirez** remembered the hands that kept him awake during the ride from Bunker Hill.

For the pilots and medics, the memories were of exhaustion and risk. Sergeant **James Moran** recalled the flashlight between his teeth as he worked on six stretchers. Captain **Oliver Johnson** remembered muscling a heavy Chickasaw into a clearing under fire. These were not isolated acts but the daily rhythm of a system that worked.

The Foundation

By the end of the Korean War, helicopter evacuation had evolved from an ad hoc measure to a well-established doctrine. The Sioux had proven the idea, the Chickasaw had carried the load, and the men of the MASH had shown what could be done when speed met surgery.

The Korean War ended in a stalemate. But out of its frozen valleys and scarred ridges came the foundation for one of the most enduring legacies of modern war — the belief that no man would be left to die on the battlefield if a helicopter could reach him.

The next generation would carry forward that creed. In Vietnam, it would be spoken in a single word over radios from rice paddies and jungles: Dustoff.

Chapter 2 Notes

- Primary Sources:

    - U.S. Army Medical Service, *Medical Support in the Korean War* (Office of the Surgeon General, 1961).

    - After Action Reports, 49th Medical Helicopter Detachment, 1951–1953.

    - U.S. Marine Corps Squadron VMO-6 records, 1952.

    - U.S. Navy Silver Star Citation, LTJG John Kevin Kane, 1951.

- Official Histories:

    - U.S. Army Center of Military History, *Korea*

1950–1953.

○ Roy E. Appleman, *South to the Naktong, North to the Yalu* (CMH, 1961).

○ Billy C. Mossman, *Ebb and Flow: November 1950–July 1951* (CMH, 1990).

○ U.S. Marine Corps History Division, *Marines in the Korean War Commemorative Series.*

• Published Works:

○ Edwin P. Hoyt, *The Chosin Reservoir Campaign* (1980).

○ James A. Huston, *Outposts Korea: The Story of the Marine Corps Helicopter Pilots* (1988).

○ Edward Hymoff, *The Helicopter Pilots* (1966).

○ T.R. Fehrenbach, *This Kind of War* (1963).

• Veteran Accounts:

○ Oral histories of Lt. Leonard B. Reuter, Capt Frank A. Carrington, and Capt Oliver L. Johnson, collected in USMC VMO-6 interviews (Marine Corps University Archives).

○ Surgeon Colonel Albert J. Glass and Dr. Paul W.

Crane, recollections preserved in U.S. Army Medical Department archives.

- Interviews with Army nurse Lt Mary Alice Hill, 8055th MASH.

- Personal recollections of medics like Staff Sergeant Harold K. Bowers and corpsmen like Sergeant James Moran, collected by the Dustoff Association and the Korean War Veterans Digital Memorial.

• Clarification:

- Most helicopter crews in Korea received the Distinguished Flying Cross or the Air Medal. Higher valor decorations, such as the Distinguished Service Cross, Navy Cross, or Medal of Honor, were rarely awarded to medevac pilots until the Vietnam War.

CHAPTER THREE

Charles L. Kelly: "When I Have Your Wounded"

Early Life and Career

Charles Luke Kelly came into the world on April 10, 1925, in Madison County, Georgia. He was one of seven children, raised in a household where there was little money and plenty of hard work. His father farmed, his mother kept the home together, and every child pulled their share. Neighbors later said Kelly had a wiry frame and restless energy, always fiddling with tools or farm machinery. That mechanical streak and determination to fix what was broken never left him.

Like many boys who grew up in the Depression, he itched to get out. World War II gave him that chance. In 1943, he joined the Army at the age of eighteen. He was sent to the infantry and by 1944 found himself overseas. Records indicate that he was in Europe during the final stages of the war. He was young, but combat hardened him quickly. His commanders marked him as a man with drive, stubborn and fearless under pressure. He returned from Europe with a soldier's bearing and a decision to stay in the Army for good.

The Korean War came five years later. Kelly did not yet fly, but he was in the Army long enough to see what the new helicopter could do. At the Chosin Reservoir and on the ridges of the Imjin, the H-13 Sioux and H-19 Chickasaw carried men out when trucks could not move. Kelly, still a ground officer, filed that memory away. He understood early that helicopters would change warfare, not as curiosities but as lifelines.

A Soldier Becomes an Aviator

By the mid-1950s, Kelly transferred into the new Army Aviation branch. He learned to fly both fixed-wing and rotary aircraft. He was not a natural pilot, but he was relentless. He drilled every maneuver, read every manual, and insisted on precision. Some who trained with him remembered his temper when things went wrong. He earned the respect of instructors because he refused to quit until he had mastered the machine.

His fellow officers noticed his habits. He inspected the aircraft down to the smallest bolt. He demanded clean logbooks. He insisted on sharp uniforms and strict discipline. To some, he was too hard, but to others, he was exactly the kind of officer they wanted on their side.

By 1963, Kelly had attained the rank of major and was known across Army Aviation as a no-nonsense officer who consistently delivered results. He had flown nearly every type of helicopter in the inventory, from the Sioux to the Chickasaw to the new UH-1 Huey. His path put him in line for command, and when the Army needed a strong hand to lead its air ambulance unit in Vietnam, Kelly's name rose to the top.

Reputation

Kelly's reputation was complicated. He could be blunt, even abrasive. One pilot, **Captain James W. "Stretch" Garrison**, later said, "Kelly scared the hell out of some people, but you knew he would never ask you to do something he wouldn't do himself." That kind of leadership built loyalty.

Another officer, **Captain Don Darius**, remembered Kelly's inspections. "He checked everything, and I mean everything. If a logbook was sloppy, or if a rotor blade was not wiped down, you would have heard about it. But when the time came to fly into fire, you wanted him leading."

Kelly's men called him "Luke the Nuke," a mix of his middle name and his explosive temper. Yet even those who clashed with him admitted he cared deeply for the soldiers who depended on him. He carried the belief that a commander's duty was to share every risk with his men.

Vietnam Beckons

By 1964, the United States was no longer just advising in Vietnam. American helicopters were flying combat support missions daily, especially in the Mekong Delta. The terrain of

Vietnam, jungles, rivers, rice paddies, and villages cut off by canals- made trucks slow and treacherous. Helicopters were the only means of transporting the wounded quickly enough to save lives.

The **57th Medical Detachment (Helicopter Ambulance)** had been the first of its kind to arrive, deploying to Tan Son Nhut Air Base near Saigon in April 1962. By the summer of 1964, it was operating from **Vinh Long**, right in the middle of the Delta. The unit had taken the call sign "Dustoff," borrowed from the clouds of dust their helicopters threw up when they landed. Soldiers on the ground quickly came to know that Dustoff meant rescue.

Kelly took command of the 57th that year. He arrived in Vietnam with his reputation already established, and within weeks, the name "Mr. Dustoff" began to follow him. He set the tone quickly. He demanded discipline, personally checked every helicopter, and made it clear that their mission was sacred.

The Huey Arrives

By the time Kelly took command, the unit was flying the **UH-1 Huey**. The Huey was bigger and faster than the Chickasaw, and it could lift a full load of patients with room for a medic to work on them. Kelly saw the Huey as the tool that would let Dustoff meet its full promise. He pushed the aircraft hard, flying deep into contested zones where even armed gunships hesitated.

His crews remembered how he would grab the radio and answer calls himself. If a platoon was hit in a rice paddy or a village was pinned down, Kelly was already airborne before anyone could stop him. "You will go where the wounded are," he told his men, "Because if you don't, they will die. That is the only measure of our work."

By mid-1964, soldiers across the Delta spoke his name. They knew that when the radio crackled with "Dustoff inbound," Kelly was likely in the cockpit. He had carried the lessons of Korea and fused them with the speed of the Huey. He had turned helicopter evacuation from a system into a creed. And for the men bleeding in the paddies, that creed had one face, Major Charles L. Kelly.

"When I Have Your Wounded"

When Major Charles L. Kelly took command of the 57th Medical Detachment at Vinh Long, he inherited more than one unit. He inherited a debate. Other aviation commanders wanted the Hueys armed for medevac. They argued that flying into hot landing zones without guns was suicide. Kelly refused.

To him, the red cross painted on the doors meant something. He believed it brought trust from every soldier in the field. If they saw Dustoff, they knew it was there for them and not for a fight. He told his superiors plainly that arming his aircraft would destroy that trust.

His stance did not make him popular with everyone. Some officers said he was reckless, that he was risking lives by flying unarmed into contested zones. Kelly shot back that his mission was not to kill but to save. "If we are going to arm these ships," he told a fellow officer, "We might as well take the red crosses off."

The Creed

Kelly drove that message into his crews. He drilled it into them at every briefing and on every flight. His pilots and medics knew that the unit's job was to go where the wounded were, no matter the fire.

Captain James W. "Stretch" Garrison, who flew under him, remembered one briefing when Kelly slammed his hand on the map and said, "If you are not willing to go in, you are in the wrong outfit." No one walked out.

Figure 1. Photo of MAJ Charles L. Kelly taken in the Republic of Vietnam in early 1964. Photo credited to a member of the United States Army (public domain).

The soldiers on the ground came to believe in that creed. In the Mekong Delta in 1964, platoons pinned in hamlets or rice paddies knew that Dustoff would come. They called for it because they trusted it would arrive, regardless of whether there were guns or not.

Kelly's Style

Kelly lived what he preached. He flew into zones others avoided, hovering in paddies while bullets snapped past the rotors, lifting men out of ditches under mortar fire. He carried them to hospitals at Vinh Long, Can Tho, and Saigon, and then he turned the Huey around and went back for more.

His crews both respected and feared him. He inspected their aircraft, barked at sloppy work, and expected them to fly the same way he did. Some thought he was too hard. Others admitted they flew better under him. What no one denied was that Kelly would always be the first bird in and the last bird out.

The Last Mission

On July 1, 1964, Kelly launched on what would be his final flight. A unit near Vinh Long had called for evacuation under fire. The landing zone was hot. He was advised to hold until the area was secure. Kelly refused. "When I have your wounded," he told them, "I will leave."

He brought the Huey in low. The fire was heavy, but he hovered steadily while his crew pulled the wounded aboard. Witnesses recalled that he never wavered. As the last man was loaded, a single round struck Kelly in the chest, killing him instantly. His Huey staggered, but his copilot and crew chief flew it out. Every wounded man on board survived.

Kelly was forty years old. He was the first Dustoff commander to be killed in Vietnam. His death rocked the 57th, but his words stayed.

The Motto

"When I have your wounded" became more than a line. It became the creed of every Dustoff crew in Vietnam. Pilots repeated it to each other. Medics wrote it in letters home. Soldiers in the field came to know it as a promise.

Captain Don Darius, one of Kelly's pilots, said later, "Kelly left us no choice. He made it clear that if there were wounded on the ground, we were going in. His death did not end that. It burned it into us."

The call sign "Dustoff" was already well-known across the Mekong Delta by 1964. After Kelly's death, it became legendary. To American soldiers, Dustoff meant help, hope, and a commander who had given his life to prove that no man would be left behind if a helicopter could reach him.

The Birth of the Dustoff Doctrine

By 1964, after the death of Major Charles L. Kelly, the word *Dustoff* meant more than a radio call sign. It had become a promise. Soldiers in the paddies and jungles of Vietnam learned that if they called for Dustoff, help would come, no matter the fire, the weather, or the risk.

Vietnam's Landscape

Vietnam demanded helicopters in a way no war had before. The Mekong Delta was a patchwork of rivers, canals, rice paddies, and villages connected by dikes and trails. Trucks bogged down or were unable to pass at all. Monsoon rains turned roads into rivers. The jungles that blanketed the Central Highlands made stretcher carries a nightmare. A man bleeding in a clearing could not wait for ground transport. He needed a helicopter within minutes.

The helicopter was no longer an experiment as it had been in Korea. It was the centerpiece of the evacuation. The **UH-1 Huey**, with its speed, power, and room for multiple patients, became the machine that tied together the battlefield and the hospital.

The 57th Medical Detachment

The unit at the heart of it all was the **57th Medical Detachment (Helicopter Ambulance)**. Originally deployed to Tan Son Nhut Air Base near Saigon in 1962, it later operated from **Vinh Long** in the Delta. The 57th had only about 40 officers and enlisted men, with a dozen Hueys on its rolls, but it carried a load far beyond its size.

After Kelly's death in July 1964, command passed to **Captain John W. "Jack" Estes**, who had been one of Kelly's pilots. Estes carried forward Kelly's creed without hesitation. "Kelly set the tone," Estes said years later. "He gave us the words we lived by, and no one thought about changing them."

The 57th flew missions across the Delta every day. A normal day meant 20 or more sorties, some of which were into contested zones. Pilots often flew until nightfall, crews snatching food in the cockpit between lifts.

Doctrine Takes Shape

By late 1964, procedures were standardized. Aid stations and infantry units marked landing zones with smoke grenades, typically using a yellow color. Radios were tied directly to the ground by the helicopter. Dustoff pilots checked in, confirmed wind and landing direction, and dropped in.

Triage became routine. Medics on the ground sorted casualties, loading the most critical first. Dustoff crews strapped them inside, and the flight medic went to work. IVs were started in the air. Dressings were tightened while the Huey shuddered under fire.

Hospitals at Saigon, Can Tho, and other bases built their rhythm around helicopters. Surgeons learned to expect waves of casualties in clusters. The time between wound and surgery, once counted in hours or days, was now often less than an hour.

The doctrine was not initially written in manuals. It was written in blood and sweat, as Dustoff pilots and medics learned what worked. But within a year, the Army codified those practices. Kelly's words — "When I have your wounded" — were not only a motto. They became policy.

Pilots and Crews

The names of the early Dustoff pilots carried weight. **Captain James W. "Stretch" Garrison** was one of them. Tall and lanky, Garrison flew dozens of missions with Kelly and Estes. He later recalled nights when crews flew until the fuel drums were dry, landing at Vinh Long only long enough to refuel and take off again.

Captain Don Darius, another of Kelly's pilots, remembered the fear of flying unarmed into hot zones. "You heard the rounds hit, you saw tracers, and you went in anyway. That was the job. Kelly drilled that into us."

Crew chiefs and medics also carried the creed. **Specialist 5 Charles Kelly Jr.,** no relation to Major Kelly, served as a crew chief in the 57th. He later described crawling across the cabin floor during flights to tighten straps on stretchers while bullets punched holes in the skin of the Huey. "You did not think about the fire," he said. "You thought about keeping them inside until you landed."

Soldiers on the Ground

For soldiers in the field, Dustoff was salvation. Letters and diaries from 1964 and 1965 are full of references to the sound of Huey rotors.

Private First-Class Robert Lane, 25th Infantry Division, wrote to his parents in late 1964: "When the chopper comes in and you see the red cross, you know you're going to make it. Dustoff doesn't leave anyone."

Sergeant William Travis, wounded near the Plain of Reeds, remembered hearing the call sign on the radio. "When I heard 'Dustoff inbound,' I knew I had a chance. I passed out after that. When I woke up, I was in Saigon with bandages on my chest."

Statistics as Story

The numbers climbed quickly. By the end of 1964, Dustoff crews had already evacuated more than **15,000 casualties.** In

1965, as American. troop numbers increased, Dustoff units flew thousands of sorties every month. Some crews averaged **30 missions a week.**

The tempo wore men down. Pilots lost weight, medics carried permanent scars from what they saw, and maintenance crews worked around the clock. But the survival rates proved the system worked. Army surgeons in Vietnam reported survival rates for the seriously wounded at levels far beyond Korea.

Everything tied back to Kelly. His refusal to arm Dustoff helicopters remained policy. His words — "When I have your wounded" — were repeated at every briefing, written into reports, and remembered by soldiers who never met him.

Captain Estes said later, "Kelly gave us the line. We lived by it. That is all there was to it."

Dustoff became one of the most recognized call signs in Vietnam. To the infantry and Marines, it meant help was on the way. To pilots and medics, it meant risk, fatigue, and danger, but also the knowledge that they were saving lives every day.

A New Era

By the time American combat units arrived in force in 1965, Dustoff was already fully established. The 57th Medical Detachment had laid the groundwork, and new units arriving from the United States adopted its procedures wholesale.

The doctrine was simple but powerful: you go in unarmed; you pick up the wounded; the medics will treat them in the air and deliver them fast to a medical facility.. It was born in Korea, forged by Kelly, and proven by the men who followed him.

From the rice paddies of the Delta to the jungles of the Central Highlands, Dustoff became a creed. It was written in the lives of the men who flew and the men they saved.

Major Charles L. Kelly was buried at **Arlington National Cemetery** on July 10, 1964. His wife, three children, and a line of fellow soldiers stood among the white headstones as the rifles cracked in salute. For the men of the **57th Medical Detachment**, it was a moment that felt both like the end of something and the beginning of something larger.

After Kelly's death, the command passed to **Captain John W. "Jack" Estes**. Estes had flown with Kelly and knew that nothing about the mission could change. In his first briefing after taking command, he told his crews, "We will honor him the only way that counts. We will fly the way he taught us. We will go where the wounded are."

From Vinh Long, the 57th kept flying. Hueys often returned shot up, skin torn, and panels rattling. Mechanics patched what they could under lantern light, sometimes working until morning, and the ships lifted again as soon as fuel and oil were in. Crew chiefs scrubbed blood off the cabin floors with rags before the next call came. Medics spoke of hands trembling from fatigue, but still they climbed back in. They all knew they were flying under a standard their commander had set, and they were not about to let it fade.

The Creed Becomes Tradition

The pilots who had flown under Kelly never forgot him. **Captain James W. "Stretch" Garrison** said, "We were stunned when he was killed, but it left us with no choice. Kelly had told us what Dustoff was. We could not back away from it."

Captain Don Darius, another pilot, put it more bluntly. "Kelly didn't give us an option. He showed us what Dustoff meant. After that, it was ours to live up to."

Enlisted men remembered him as hard, even unforgiving, but they also remembered that he never spared himself. **Specialist 5 Charles Kelly Jr.** recalled, "He was tough as nails. He chewed us out, sure. But he was always in the cockpit, taking the same fire, we took. That counted more than anything."

Within weeks of his death, his line — *"When I have your wounded"* — was being repeated across Vietnam. It was not written on posters or manuals. It spread the way soldiers' talk always spreads, passed from crew to crew, from platoons in the paddies to surgeons in the tents.

Kelly's Awards

For his final mission at Vinh Long, Kelly was posthumously awarded the **Distinguished Service Cross**, the Army's second-highest decoration for valor. He had already earned the **Legion of Merit**, the **Air Medal**, and the **Army Commendation Medal** during his career. The DSC was recognition not only of one flight but of the creed he had given to his men.

Distinguished Service Cross Citation – Major Charles L. Kelly

The President of the United States of America, authorized by Act of Congress, July 25, 1963, takes pride in presenting the Distinguished Service Cross (Posthumously) to Major Charles Luke

Kelly (ASN: 0-96292), United States Army, for extraordinary heroism in connection with military operations involving conflict with an armed hostile force in the Republic of Vietnam, while serving with the 57th Medical Detachment (Helicopter Ambulance), 67th Medical Group, United States Army Vietnam. On 1 July 1964, Major Kelly was serving as aircraft commander of a UH-1 helicopter ambulance on a mission to evacuate wounded personnel from a United States advisory unit under hostile fire near Vinh Long, Republic of Vietnam. Disregarding repeated warnings that the area was too dangerous for evacuation, Major Kelly insisted on attempting the mission. While hovering under intense small-arms fire to load wounded personnel, Major Kelly was mortally wounded. His gallant decision to fly into an extremely dangerous situation and his refusal to depart until the wounded were aboard resulted in the evacuation of several personnel who otherwise would have perished. Major Kelly's conspicuous heroism and supreme devotion to duty were in keeping with the highest traditions of the military service and reflect great credit upon himself, his unit, and the United States Army.

Impact Beyond the 57th

Word of Kelly's death spread across Army Aviation. In Washington, senior leaders cited him as the embodiment of Army Aviation's role in Vietnam. In Vietnam itself, his death hardened the identity of every Dustoff crew that followed. New pilots arriving in the country were told Kelly's story before they ever flew their first mission.

Colonel Patrick H. Brady, who would later command Dustoff units and earn the Medal of Honor, said years later, "We all flew with Kelly's words in our ears. They became the standard. They were why we went in."

By 1965, Dustoff had expanded far beyond the 57th. New medical detachments arrived with dozens of Hueys. The red cross painted on their doors was a symbol understood across the country. Infantry and Marines tuned their radios to call for Dustoff, just as their fathers had once called for artillery.

The Numbers Tell the Story

By the war's end, Dustoff crews had evacuated over **900,000 casualties.** Survival rates soared to more than ninety percent for those who reached a Dustoff helicopter alive. Historians and Army medical officials alike credited Kelly with setting the uncompromising standard that made those numbers possible.

It was not just about machines or tactics. It was about the will to go where the wounded were. That was Kelly's gift to the men who followed.

Mr. Dustoff Remembered

Kelly's men remembered him not as a saint but as a commander who was hard, demanding, and uncompromising. They also remembered that he lived the mission he demanded of them.

Captain Estes summed it up best. "Kelly was Dustoff. The rest of us just carried it on."

That was his legacy. Every Huey with a red cross that flew into fire carried its creed. Every soldier pulled out of a rice paddy or a jungle clearing owed something to the standard he set.

For generations of Dustoff pilots, the measure of the mission
remained his words: *When I have your wounded.*

   You never asked if it was safe.
You just came. Sometimes in daylight,
sometimes in the black,
rotors chopping dust and smoke,
rounds snapping at your tail.
   We remember the sound first —
that low thump rolling in,
the kind of sound
that told a man he might see home.
   You landed where no one else would.
Fields, alleys, rice paddies,
roads wired with bombs.
Didn't matter.
If someone was down,
you found a way.
   The back of your bird
was more than metal.
It was hope.
It was a promise.
   And you kept it,
again and again,
until some of you didn't make it back.
   That's the part we carry still.

When I have your wounded —
that wasn't just a line.
It was you.
It will always be you.

Chapter Notes:

Primary Sources:

Headquarters, United States Army Vietnam, General Orders No. 252 (1965) awarding the **Distinguished Service Cross** posthumously to Major Charles L. Kelly.

57th Medical Detachment unit records, 1962–1964, U.S. Army Aviation Museum, Fort Novosel, Alabama.

U.S. Army Aviation Digest, "Major Charles L. Kelly and the Dustoff Tradition," (1965).

Official Histories:

Office of the Surgeon General, U.S. Army, *Medical Evacuation in Vietnam* (Washington, 1970).

U.S. Army Center of Military History, *Vietnam Studies: Medical Support of the U.S. Army in Vietnam,1965–1970*, by Major General Spurgeon Neel.

U.S. Army Center of Military History, *Vietnam Studies: Air-mobility 1961–1971*, by Major General James K. McDonough.

Published Works:

Peter Dorland and James Nanney, *Dust Off: Army Aeromedical Evacuation, 1961–1991* (U.S. Army Center of Military History, 1991).

Peter Davis, *Vietnam at War: The History, 1946–1975* (1983).

Philip D. Chinnery, *Any Time, Any Place: Fifty Years of the USAF Air Rescue Service* (1987).

Veteran Accounts:

Interviews with Captain John W. "Jack" Estes, Captain James W. "Stretch" Garrison, and Captain Don Darius, collected in the U.S. Army Aviation Museum oral history archives.

Letters and recollections of crew chiefs and medics from the 57th Medical Detachment, preserved by the Dustoff Association.

Remarks of Colonel Patrick H Brady, Medal of Honor recipient, on Kelly's legacy, recorded in Dustoff Association newsletters (1980s).

Clarification:

Major Charles L. Kelly's awards included the **Distinguished Service Cross (posthumous), Legion of Merit, Air Medal, and Army Commendation Medal.** His DSC citation for July 1, 1964, recognizes his extraordinary heroism in evacuating wounded under fire near Vinh Long, Vietnam.

Kelly's death marked the first combat loss of a Dustoff commander in Vietnam. His last words, *"When I have your wounded,"* became the enduring motto of Army aeromedical evacuation.

Chapter Four

The 57th Medical Detachment: "Dust Off"

When Major Charles L. Kelly was killed at Vinh Long in July 1964, the 57th Medical Detachment lost its commander, but its identity did not change. His last words, "When I have your wounded," fused with the call sign the unit already carried. From that day forward, "Dust Off" was more than a voice on the radio; it was a covenant. For the rifleman lying in a jungle clearing, it meant rescue. For the pilots, crew chiefs, and medics, it meant duty, whatever the odds.

The 57th Medical Detachment, Helicopter Ambulance, came to Vietnam in 1962 with a handful of UH-1A Hueys and a simple mission: to carry the wounded from the battlefield to surgical care as quickly as the weather and enemy would allow.

Origins of the 57th

Early crews operated from Tan Son Nhut, near Saigon, then split into small operating sections to cover more ground. Detachment A settled in Soc Trang, in the Mekong Delta, while Detachment B moved to Nha Trang on the central coast. Later, a forward section was deployed to Pleiku in the Central Highlands. The dispersion shortened the clock; a wounded man may need to wait from an hour to what might now be fifteen minutes away.

The first model Hueys were underpowered in tropical heat; three or four litters were about the limit if the fuel load was heavy and the air was thin. The aircraft were unarmed; their most visible protection was the red cross on the doors. The call sign "Dust Off" was chosen early, a short, precise phrase that fit what Hueys did in unimproved landing zones; they kicked up choking clouds of dirt. The name spread through adviser nets and Vietnamese units; by the end of 1963, "Dust Off" meant the deed as much as the machine.

What a crew looked like

A Dust Off crew was four people: two pilots, a crew chief, and a medic. The aircraft commander flew from the right seat, the copilot handled radios, navigation, and engine instruments from the left. The crew chief lived with the aircraft; he pulled chocks, watched the tail, called obstacles, and kept the bird flying with long hours after the missions were done. The medic was the difference between transport and rescue; he ran triage on the skid, stopped the obvious bleeding, secured airways, began fluids when possible, and kept men alive until surgery.

The medic's kit was compact yet brutal, containing battle dressings, tourniquets, IV sets, a limited stock of morphine syrettes, plastic airways, splints, tape, shears, and a stethoscope that often hung useless in the rotor wash and gunfire. In the cabin, there were litter straps and handholds everywhere. The medic learned to brace with a knee and work by feel when the Huey pitched.

How a mission worked

The radio call could come at any hour. A ground unit sent a nine-line request through the medical channel, specifying the unit, location, number, and priority of casualties, security of the landing zone, special equipment needed, marking signal, patient nationality, and status. The pilot copied the grid, the crew chief threw open the doors, the medic hauled the kit aboard, and the Huey was moving before the second hand made a circle.

The approach was dictated by terrain and fire. If the landing zone was open and quiet, the pilot flew a quick straight-in, settled the skids, and the crew worked in seconds. If the zone was hot, the pilot jinked, used tree lines for cover, curved in low with a tight flare just long enough to touch, then went to power. Ground troops popped colored smoke to mark the pickup; the pilot had to confirm the color because the enemy sometimes threw their own smoke to lure aircraft to the wrong spot. At night, chem-lights marked paths, flares turned the jungle into a false daylight that made distance hard to judge.

Dust Off often flew without a gunship escort because delays cost lives. Artillery suppression and fast movers covered what they could. The Red Cross was supposed to mean protection; everyone learned quickly that it did not. If the zone could not take a landing, the crew used the rescue hoist, a slow and exposed method that demanded tight coordination between the pilot and the crew chief. The hoist saved lives from the tree line and ditch; it also drew fire like a magnet.

Life around the flight line

Most of the 57th lived a few steps from the revetments. Hooches were made of sheet metal and wood; inside, there were cots, a fan (if you were lucky), and a small shelf for letters. Boots stayed at the foot of the bed, laces loose for speed. Meals were quick; there were days when crews ate C-rations on the skid between sorties. Maintenance went on long into the night; bullet holes were patched with speed tape and sheet aluminum, hydraulic lines were swapped by lantern, and rotor blades were hammered for balance. Spare parts were always in short supply; cannibalization was a common practice. One grounded Huey became the donor to keep three others alive.

Pilots logged two hundred to two hundred fifty hours in heavy months. The noise stayed in the bones; men heard rotor slap even in sleep. The smell of fuel, oil, and blood permeated the cabin's paint. Crews kept a simple superstition: clean the cabin at the end of the day, do not carry yesterday's blood into tomorrow's flight.

Carrying Kelly's creed

When Major Kelly took command in early 1964, he gave the 57th something more than orders; he gave it a tone. The unit already had experience; he supplied an edge. Go when called, hold the skids as long as it takes, do not leave a man who can live. His death fixed that ethic. New pilots learned procedures from manuals and checklists; they knew what Dust Off meant from the older crews who could still hear Kelly's voice, a refusal to abandon the wounded. By the time large American combat formations arrived in 1965, Dust Off had a personality that infantrymen recognized instantly. It was aggressive, it was predictable, and it came.

The road into War Zone C

In December 1967, the 25th Infantry Division opened Operation Yellowstone in Tây Ninh Province, the target, War Zone C, a long-held Viet Cong stronghold near the Cambodian border. It was a country of dark jungle, rubber plantations, termite mounds, and spider-web trails that ran to sanctuaries beyond reach. The mission was to break the enemy's base areas before they could shape the new year on their own terms.

For Dust Off, Yellowstone meant constant movement, Katum and Tay Ninh West, Dau Tieng, and small firebases cut out of the bush. Every battalion knew the medevac frequency. The 57th's aircraft came and went over the canopy, a green pulse that meant the casualty chain was working. Surgeons at clearing stations kept scapulas ready; Dust Off turned distance into minutes, and surgeons turned those minutes into lives saved.

Katum, 2 January 1968

A rifle company north of Katum found itself in a shaped ambush just after midday, mortars first, then machine guns from a tree line the patrol had skirted too closely. Men went down into the grass; the company broke into small groups, returning fire as they beat back the brush. The call went out for Dust Off. The first Huey dropped in fast and light, four litters aboard, the medic bracing in the door while the crew chief hauled by web gear and belt. As the aircraft lifted, green tracers walked across the tail. The pilot felt the tail boom buck and kept climbing.

There was no time to count holes. The second Huey was already curving in, making a shallow S-turn to allow the gunners to adjust, followed by a flare, as the skids touched down, and the crew moved like a drill they had practiced a hundred times. A man with a sucking chest wound came aboard first. The medic sealed the hole with a dressing and plastic wrap, taped it down tight, started his fluids, and shouted the number of patients forward to the pilot so the clearing station could be warned. The aircraft began to rise and settled again as another casualty was shoved into the cabin. The pilot held until he felt the crew chief slap the bulkhead, signaling the power to be pulled.

Trip followed trip, two aircraft shuttling from the same clearing to the aid station and back, casualties stacked in order of need, urgent, priority, routine. The route was never the same twice; the pilots varied their altitude and track. Artillery fired suppression on the tree line. The air was a web of moving pieces, all of it meant to carve a five-minute window where a red cross could live. By evening, more than forty men had been carried off that grass. Both aircraft came home with fuel cells patched, blades nicked and scabbed, paint burned away by muzzle flashes. No one counted the holes until the rotors stopped.

Fire Support Base Burt, 8–9 January 1968

The enemy came for FSB Burt just after dark, a regimental-sized force that probed, then pressed, then tried to smother the base under weight. Mortars lifted, rockets struck the gun pits, machine guns skimmed the parapets, and a steady assault moved from trench to trench. Infantrymen fired until the barrels burned their hands. Flares went up from the base and from circling aircraft; the night turned a dirty white, flattening depth and making the world feel close.

Dust Off called the perimeter, confirmed the mark, and came in. The first Huey settled inside the wire while the guns still spoke. Crew chief and medic moved at a crouch, dragging casualties by web straps and lifting litters two men to a side. The pilot kept a steady hover until the weight indicated the cabin was complete; the crew chief slapped the bulkhead, and the ship nosed into a shallow climb, tracers lifting with it like angry vines. Outside the wire, red lines crossed the dark. Inside the cabin, the medic did the same work he always did: cut pant legs, find and stop the bleeding, secure an airway, push fluid, talk into a man's ear, and tell him he was not alone.

There were too many to lift in one turn. The first aircraft cleared the base and ran for the clearing station; the second was already touching down. Ground guides used flashlights cupped in their hands to shield the beams, while flares hissed and drifted, the air stank of cordite, diesel, and blood. The battle raged into dawn, Dust Off aircraft shuttled without pause, and by the time the horizon turned, dozens of men had reached surgeons who had been waiting with their sleeves rolled up.

One crew came back with a hydraulic line leaking into the cabin. The crew chief kept a rag tied tight around the fitting while the pilot eased the aircraft down. Another returned with a rotor blade scarred; the maintenance chief swapped blades by lantern light and a metal yardstick while the flight medic slept on the concrete with his head on a helmet bag. At first light, the same aircraft lifted again.

A participant in the battle for Burt was a young Oliver Stone. Survivors of the battle often remarked that the last scenes in the movie, Platoon, with the enemy trying to overrun the American positions, were indicative of what Stone had experienced at Burt.

What does the grind cost

Yellowstone pushed crews to the edge. Pilots logged two hundred to two hundred fifty hours that month; fatigue made hands shake during preflight, eyes burned in the glare of noon, and the false light of flares at night. Medics wrote home about the smell that would not wash out: hydraulic fluid, JP-4, and human blood. Crew chiefs developed the habit of counting litter straps by touch, once for the lift, once again in the air, and once more before landing, as a safety measure.

Maintenance was a second battle. Dust and sand ate away at engines and gears, monsoon water seeped into everything, and corrosion worked even when the aircraft was at rest. Supply was always just short of enough. The shop kept a small museum of damaged parts to teach new hands what blast and bullet did to aluminum, haste, and luck. The rule was simple: if a part could be made to work safely, it was back on the aircraft by nightfall.

Why soldiers trusted the call

The infantryman measured Dust Off not in statistics, but in the minutes it took to hear blades after the radio call. He measured it in the sight of a red cross settling into a field where he would not have risked a jeep. He measured it in the way his wounded friend's fingers regained color before the helicopter cleared the tree line. He told the new man in the squad, 'If you get hit, keep breathing and speak up; Dust Off will come.'

Commanders measured Dust Off differently, but with equal respect. They saw the effect on morale; troops pushed further when they believed the chain of evacuation would hold. They saw it in after-action reports, where the time from injury to surgery shrank to a fraction of what previous wars had endured. Generals spoke about contributions and systems; the line soldiers used simpler words. We do not pray for miracles out here; we pray for Dust Off.

What made the 57th the seed unit

The 57th trained replacements from other medical detachments, and it wrote standing procedures that others copied. It proved that a fixed call sign and a medical chain under medical control saved lives. The crews taught habits that are now doctrine, confirming the smoke color to defeat deception, varying routes and altitudes to break patterns, coordinating artillery suppression and timing for the landing, landing even when the zone is tight if the ground unit can secure the approach, lifting the worst first, then coming back. They taught that a helicopter could be both fragile and relentless, unarmored and unstoppable, if the crew believed the job was worth their lives.

The lesson of Yellowstone was not a slogan; it was a set of practices hammered into muscle memory. The 57th was not the only unit to fly medevac in those months, but it remained the emblem in the minds of many who fought there, the voice that answered in clear English through static, Dust Off inbound, mark the zone, we see your smoke, we are coming in.

The arithmetic of mercy

By the end of the Vietnam War, Dust Off helicopters had carried close to nine hundred thousand patients, Americans and Vietnamese, soldiers and civilians. More than two hundred medevac crewmen were killed while doing it. Survival rates for those who reached a Dust Off aircraft were unlike any previous war; the speed of lift and the presence of trained medics in the cabin changed the curve. Those numbers are the dry side of the story; the living side is simpler: thousands of men who went home because a red cross appeared over a tree line when it mattered.

The handoff to the rest of the war

Operation Yellowstone closed with the enemy still present and the war still vast, but the doctrine was set. Dust Off had proven that a predictable call sign, a dedicated unit, and crews trained to land under fire could change outcomes across a theater. The 57th's veterans moved to other detachments and carried the habits with them; the call sign remained fixed so a frightened voice could always find the right name to say; the surgeons kept their lamps lit because they trusted the thump of Huey blades to bring them work and to send men back into the world.

Major Kelly had spoken a promise in the Mekong Delta in 1964. The 57th kept it in War Zone C from 1967 to 1968, at Katum, Fire Support Base Burt, and in small locations that never appeared on a map. The men who flew the missions would not have called themselves heroes; they would have said they were doing their job. The men they carried out used different words. They called them Dust Off.

Dust Off

They called, and you came, blades beating the air,
a cross on your doors, no guns on your sides,
only the promise of hope.
You came through tracer fire, through smoke and dust,
settling skids in places no one thought you could land.
Hands reached up from the dirt, hands slick with blood,
and you pulled them aboard, one more life lifted from the
edge of silence.

You carried the broken, the frightened, the dying.
Your decks stained red, your medics bent low,
whispering, "Hold on, we'll get you home."
Many of you did not return. But still the blades turned,
and still the promise held: When I have your wounded.
That was your creed. That was your gift. That was Dust
Off

Notes:

☐ **Primary Sources:** Unit histories of the 57th Medical Detachment (Air Ambulance), 1964–1968, U.S. Army Center of Military History.

☐ **Operational Records:** "After Action Report, Operation Yellowstone," January–February 1968, 25th Infantry Division Archives.

☐ **Published Works:** Matt Jackson, *Dust Off: Army Aeromedical Evacuation in Vietnam* (Texas A&M University Press, 2000); Peter Dorland and James Nanney, *Dust Off: Army Aeromedical Evacuation in Vietnam* (Center of Military History, 1982).

☐ **Veteran Accounts:** Oral histories collected by the Vietnam Helicopter Pilots Association, interviews with 57th Medical Detachment pilots and crewmen, 1970s–2000s.

☐ **Supplemental Material:** Articles from *Army Aviation Digest*, 1967–68, covering the development and role of helicopter ambulance units in Vietnam.

CHAPTER FIVE

Bruce Crandall: Landing Zone X-Ray

The air over the Ia Drang Valley in November 1965 shimmered with smoke and the crack of gunfire. At the edge of a small clearing called **Landing Zone X-Ray**, a UH-1 Huey dropped into the fight. Bullets snapped across the grass, striking metal and Plexiglas. In the right seat sat **Major Bruce Perry Crandall**, forty-two years old, a career Army aviator who had spent more than a decade mapping jungles, hauling supplies, and training for a new kind of war. On that day, his mission was simple and deadly: carry ammunition, water, and medical supplies into a battalion that was surrounded, and take the wounded back out through a storm of enemy fire.

The fight at X-Ray became the first major battle between U.S. Army forces and the North Vietnamese Army. For Lieutenant Colonel **Hal Moore's** 1st Battalion, 7th Cavalry Regiment, it was three days of desperate fighting. For Crandall and his wingman, **Captain Ed Freeman**, it was sixteen hours of repeated flights into hell, more than twenty-two sorties through fire that never slackened. By the time the sun set on November 14, their Hueys were riddled with holes, their crews were near exhaustion, but the men of the 7th Cavalry were still holding, supplied, reinforced, and alive.

Early Life in Washington

Bruce Perry Crandall was born on **February 17, 1933**, in Olympia, Washington. He was the oldest of three children, raised in a modest home where his father worked long hours to provide for the family, and his mother managed the household. The family did not have much in the way of money, but they had discipline and a love of sports.

Crandall grew up as a competitive and stubborn individual, traits that sometimes landed him in trouble but also marked him as a natural leader. Baseball was his first love. At Olympia High School, he pitched and played shortstop, impressing coaches with his strong arm. Friends remembered him as "the kid who hated to lose more than anyone else," a quality that stayed with him for life. There was talk of a shot at professional baseball, but the Korean War was raging, and the Army claimed him first.

At West Point, he continued to play the sport, where he was a member of the cadet team while balancing a demanding schedule of academics and military training. He graduated from the **Class of 1953** and was commissioned into the Corps of Engineers. Aviation was beginning to expand in the Army, and Crandall soon transferred into that field, drawn by the challenge of flying and the chance to be part of something new.

Survey Pilot in Remote Places

Before Vietnam, Crandall logged thousands of hours over remote wilderness. Assigned to Army mapping and survey units, he flew low and slow over the mountains and jungles of Central and South America. He carried cameras and crews who mapped rivers and terrain that had never been charted. His assignments took him over the Andes in Ecuador, into the dense forests of Peru, and across the vast reaches of the Amazon basin.

These were not routine flights. The weather in the tropics changed without warning. Thunderstorms rose in minutes, winds buffeted light aircraft, and engine failures meant forced landings in terrain where there were no roads or clearings. On one mission in Ecuador, Crandall lost an engine over a canyon and had to nurse the aircraft into a narrow gravel bar along a river, narrowly avoiding disaster. He later described such episodes as "routine hazards," but they taught him to keep calm under pressure and to think quickly in the cockpit.

Those years gave him something more valuable than flight hours. They gave him confidence that, no matter the conditions, he could safely land an aircraft and keep his crew alive. In

Vietnam, that confidence would mean the difference between landing in a hot zone or pulling away.

The Airmobile Revolution

By the early 1960s, the Army was experimenting with an idea that would revolutionize its approach to warfare: airmobile warfare. Helicopters had shown their value in Korea for evacuation and supply. The arrival of the UH-1 Huey, with its greater lift and reliability, made it possible to move entire units by air.

The **11th Air Assault Division** at Fort Benning tested the concept. Its mission was to see whether infantry could be trained and equipped to fight as an air-mobile force. When the experiment proved successful, the division was reflagged as the **1st Cavalry Division (Airmobile)**. For the first time in Army history, helicopters were not just support; they were the backbone.

Crandall took command of **Company A, 229th Assault Helicopter Battalion**, part of the 1st Cav. His company was responsible for lifting infantry battalions into battle and keeping them supplied once they were in. At Fort Benning, his men practiced flying in mass formations, setting down entire rifle companies in sequence, and coordinating with artillery and gunships to suppress landing zones. The training was intense and unrelenting. Older officers were skeptical, but younger aviators sensed they were on the front edge of something revolutionary.

In 1965, the division deployed to Vietnam and established a base at An Khe in the Central Highlands. The terrain was

rugged, the enemy aggressive, and the chance to test the airmobile concept under fire came sooner than anyone expected.

The Road to Ia Drang

By November 1965, intelligence reported large North Vietnamese units moving into the Chu Pong Massif, a mountain
range straddling the Cambodian border. The mission went to
Lieutenant Colonel Hal Moore and his 1st Battalion, 7th Cavalry. He would lead his men into the valley to find and destroy
enemy forces. Crandall's company would deliver them.

On **November 14, 1965**, before dawn, Crandall's Hueys
lifted off from Plei Me with companies of the 7th Cavalry. The
first waves landed smoothly at **Landing Zone X-Ray**, a clearing
at the base of the mountains. Soldiers ran into the grass, set up
a perimeter, and waited for reinforcements to arrive. As more
companies arrived, the enemy launched its attack. By midday,
mortars and automatic weapons raked the clearing. Hundreds
of North Vietnamese troops were encircling the Americans.
Casualties mounted quickly, and Moore's perimeter strained
under the assault.

Moore called for supplies and evacuation. Without resupply, his battalion would be overrun. Without evacuation, his
wounded would die in the dirt. The only lifeline was through
the sky, and the men answering were Bruce Crandall and Ed
Freeman.

Flying Into the Fire

Crandall volunteered to lead the resupply effort. He and
Freeman began shuttling loads of ammunition, water, and ban-

dages into the clearing. Each approach meant flying through a gauntlet of fire. Mortar rounds bracketed the zone, machine guns tracked the approaches, and every touchdown was a gamble.

The first runs carried ammunition crates. Infantrymen rushed the loads off as soon as the skids touched. Immediately, the wounded were dragged aboard. Some groaned, some were silent, blood pooling on the floor. Crew chiefs and medics worked swiftly, securing litter straps, cutting away gear, and applying pressure to wounds.

Crandall flew with precision under pressure. His Huey came in low, flared hard, and held just long enough to load before pulling away. Freeman followed close behind, splitting the fire. They repeated the pattern again and again.

Crandall's bird at LZ X-Ray
Public Domain

A battalion surgeon pleaded with them to keep coming. "If you don't bring ammo, we die. If you don't take the wounded, they die." The truth of that statement drove them back into the teeth of the fire every time.

Hour by Hour at X-Ray

The day stretched into a blur of sorties.

Morning: Crandall led the initial lifts, bringing in the companies of the 7th Cavalry. By mid-morning, casualties began piling up. His Huey came back from the first evacuation flight riddled with bullet holes. Mechanics patched what they could in minutes, wiping away hydraulic fluid, sealing holes with tape, and sending the ship back out.

Midday: Fire intensified. Enemy mortars bracketed the landing zone. Crandall and Freeman flew with their rotors barely above the treetops, using every trick they knew to reduce exposure. At one point, Crandall's Huey lifted so overloaded with wounded that the skids dragged through the grass before he could coax it into the air.

Afternoon: Exhaustion set in, but the pace did not slow. Supplies came in waves, ammunition stacked in crates beside crates of water. Wounded were evacuated by the dozen. The air inside the cabin was thick with blood, sweat, and cordite. A crew chief recalled, "You couldn't see the floor. It was just bodies and bandages. Major Crandall never flinched. He held steady until we were full, then pulled us out again."

Evening: As dusk fell, the fight raged on. Crandall and Freeman made their final sorties under failing light, guided by smoke and muzzle flashes. Their aircraft came back scarred, but they had delivered what the battalion needed to survive the night.

By day's end, they had flown more than twenty-two sorties, delivered nearly thirty tons of supplies, and evacuated over seventy wounded. Their Hueys bore the marks of the fight, but they had not faltered.

Aftermath

The Battle of Ia Drang lasted three days. Moore's battalion lost seventy-nine men killed and more than 120 wounded. The

North Vietnamese lost hundreds. The fight proved the ferocity of the enemy and the value of the airmobile concept.

Crandall and Freeman were recommended for high awards, but due to the confusion with paperwork, they received only the **Distinguished Flying Cross**. It was not until decades later, after veterans and historians pressed the case, that the recognition was upgraded. Freeman received the **Medal of Honor** in 2001. Crandall received his in 2007 in a White House ceremony. His citation credited him with "extraordinary heroism and selfless devotion to duty" and praised his actions for saving the lives of countless soldiers.

Beyond medals, Crandall was remembered for his calm presence. Pilots who flew under him said he never panicked. One recalled, "He didn't yell. He didn't lose his head. He just flew. You looked at him and thought, if he's steady, we're steady."

After Vietnam, Crandall continued to serve, flying additional tours and later commanding aviation units. He retired as a lieutenant colonel after twenty-two years in uniform. In civilian life, he remained tied to veterans, speaking at reunions and mentoring younger soldiers. He told cadets that leadership was not about speeches, but about showing up when needed.

The Measure of Ia Drang

For the infantry, Ia Drang was a baptism of fire. For aviation, it was proof that helicopters could tip the balance of battle. Without the flights of Crandall and Freeman, Hal Moore's battalion might not have survived.

Moore later wrote, "They saved our lives, all of them. They flew until they could no longer fly. They carried the wounded and brought us what we needed to hold." Reporter **Joseph Galloway**, who was at X-Ray, never forgot the sight of Hueys dropping through sheets of fire. For the wounded, the sound of rotors overhead meant survival.

Legacy

Bruce Crandall's flights at Ia Drang became part of Army legend. The story was told in Moore and Galloway's "We Were Soldiers Once... and Young," and later in the film adaptation. But his real legacy lay in the creed he lived by: the mission comes first, the wounded come first, leadership means presence.

Bruce Crandall. He would be awarded the Medal of Honor later. Public Domanin

In that, he carried forward the tradition of Charles Kelly. Dust Off was not only about helicopters; it was about faith, the faith of soldiers that someone would come. At Ia Drang, Bruce Crandall proved that faith was justified.

Citation: For conspicuous gallantry and intrepidity at the risk of his life above and beyond the call of duty: Major Bruce P. Crandall distinguished himself by extraordinary heroism as a Flight Commander in the Republic of Vietnam, while serving with Company A, 229th Assault Helicopter

Battalion, 1st Cavalry Division (Airmobile). On 14 November 1965, his flight of sixteen helicopters was lifting troops for a search and destroy mission from Plei Me, Vietnam, to Landing Zone X-Ray in the Ia Drang Valley. On the fourth troop lift, the airlift began to take enemy fire, and by the time the aircraft had refueled and returned for the next troop lift, the enemy had Landing Zone X-Ray targeted. As Major Crandall and the first eight helicopters landed to discharge troops on his fifth troop lift, his unarmed helicopter came under such intense enemy fire that the ground commander ordered the second flight of eight aircraft to abort their mission. As Major Crandall flew back to Plei Me, his base of operations, he determined that the ground commander of the besieged infantry battalion desperately needed more ammunition. Major Crandall then decided to adjust his base of operations to Artillery Firebase Falcon in order to shorten the flight distance to deliver ammunition and evacuate wounded Soldiers. While medical evacuation was not his mission, he immediately sought volunteers and with complete disregard for his own personal safety, led the two aircraft to Landing Zone X-Ray. Despite the fact that the landing zone was still under relentless enemy fire, Major Crandall landed and proceeded to supervise the loading of seriously wounded Soldiers aboard his aircraft. Major Crandall's voluntary decision to land under the most extreme fire instilled in the other pilots the will and spirit to continue to land their own aircraft, and in the ground forces the realization that they would be resupplied and that friendly wounded would be promptly evacuated. This greatly enhanced morale and the will to fight at a critical time. After his first medical evacuation, Major Crandall continued to fly into and out of the landing zone throughout the day and into the evening. That day he completed a total of 22 flights, most under intense enemy fire, retiring from the battlefield only after all possible service had been rendered to the Infantry battalion. His actions provided critical resupply of ammunition and evacuation of the wounded. Major Crandall's daring acts of bravery and courage in the face of an overwhelming and determined enemy are in keeping with the highest traditions of the military service and reflect great credit upon himself, his unit, and the United States Army.

Notes:

□ **Primary Sources:** After Action Report, 1st Battalion, 7th Cavalry Regiment, November 1965; aviation support records from the 229th Assault Helicopter Battalion, 1st Cavalry Division.

□ **Official Histories:** Harold G. Moore and Joseph L. Galloway, *We Were Soldiers Once... and Young* (Random House, 1992), which includes Crandall's role at Landing Zone X-Ray; U.S. Army Center of Military History monographs on the Pleiku Campaign, 1965.

□ **Published Works:** Shelby L. Stanton, *Vietnam Order of Battle* (Galahad Books, 1981); John M. Carland, *Stemming the Tide: May 1965 to October 1966* (CMH Vietnam Studies, 2000).

☐ **Veteran Accounts:** Oral histories of Company A, 229th Assault Helicopter Battalion, preserved in the Vietnam Helicopter Pilots Association archives; interviews with Bruce Crandall conducted in the 1990s–2000s.

☐ **Supplemental Sources:** Medal of Honor citation for Lt. Col. Bruce P. Crandall, Department of Defense, February 2007.

CHAPTER SIX

Ed Freeman "Too Tall to Fly, Too Brave to Quit"

On November 14, 1965, the clearing known as Landing Zone X-Ray resounded with the thump of Huey blades, the crack of mortars, and the sharp rip of machine-gun fire. Two helicopters kept coming back, again and again, into the storm. The first belonged to Major Bruce Crandall, already known as a bold and aggressive pilot. The second was flown by **Captain Ed W. Freeman**, a tall, quiet Mississippian whose calm hands and steady manner reassured the men in his ship.

The infantry on the ground noticed both men. To them, Crandall was fire and Freeman was steel. One commanded attention, the other radiated calm, but together they delivered what was needed most: supplies in, wounded out, lives saved.

Mississippi Roots

Edward C. Freeman was born on **November 20, 1927**, in Neely, Mississippi. He was the sixth of nine children, part of a large family that lived through the hardships of the Great Depression. His father worked where he could, farming, hauling timber, repairing tools. His mother managed the household with discipline and patience. Every child had chores before and after school.

Neighbors remembered the Freeman children as hard workers, especially Ed, who, even as a boy, was dependable and quiet. He grew tall early, so tall that by the time he reached high school, he was already over six feet. At full growth, he stood six feet four inches, earning the nickname "Too Tall." His height would later be called a disqualification when he applied for pilot training. For now, it made him a center on the basketball team and a figure impossible to miss in small-town Mississippi.

Life in rural Mississippi left its marks. Freeman loved the outdoors, spending hours hunting squirrels or fishing in local streams. He learned patience with a rod and determination with a rifle. His classmates recalled him as polite, serious, and reliable. He was not a showman, but he could be counted on. Those qualities became his signature for the rest of his life.

A Sailor in World War II

When World War II reached its climax, Freeman was still a teenager. Like many boys of his generation, he felt the call of service before finishing school. He enlisted in the **Navy** near the end of the war and found himself assigned to the **USS Cacapon**, a fleet oiler that kept the Navy's warships fueled across the Pacific.

Life aboard an oiler was a life of long, routine days and danger. Crews worked shifts around the clock, refueling carriers and destroyers at sea while both ships steamed side by side. One slip of a line or spark of static could mean fire and disaster. Freeman, not yet twenty, learned discipline and teamwork in those months. The Pacific gave him a taste of the world beyond Mississippi and a sense of belonging to something larger than himself.

The war ended before he saw combat, but he came home with experience and maturity that set him apart from his peers. He had left as a farm boy, and he returned as a sailor who knew what it meant to be part of a crew that depended on each other absolutely.

Serving his country again

In 1948, Freeman enlisted again, this time in the **Army**. He began in the Corps of Engineers, working on projects that ranged from road construction to bridge building. But aviation caught his eye. He applied for flight training, only to be told he was too tall. At six foot four, he exceeded the Army's limit for pilots. For his immediate future Freeman was a member of an engineering company and when the fighting started in Korea, his unit found itself in the midst of combat, as infantry. Freeman was serving as first sergeant, when In the opening phases of the Battle for Pork Chop Hill his leadership was tested repeatedly.

In the bitter fighting he was one of 14 men still able to fight out of the 257 that entered the fray. He received a battlefield commission to Lieutenant and was soon leading a company.

When he returned, he found that he still had a desire to fly, to be like the birds he remembered in Mississippi. At 6 foot 4, the army continued to reject his request as his height disqualified him from pilot training. Most men would have accepted the verdict. Freeman did not. He applied repeatedly until the army had raised the accepted height of applicants. In 1955, he earned his wings. The moment was a triumph over bureaucracy and doubt. He wore the nickname "Too Tall" with pride. He had been told no, and he had turned it into yes.

His early flying duties involved flying the Bird Dog, a light observation aircraft manufactured by the Cessna company. He found himself doing mapping duties throughout the world before being trained as a helicopter pilot in the expanding airmobile forces of the army.

Into the Airmobile Age

By the early 1960s, helicopters were reshaping the Army. Freeman transitioned into rotary-wing flying and mastered the UH-1 Huey. He was assigned to the **11th Air Assault Division** at Fort Benning, the Army's test unit for air mobility. The concept was bold: move entire battalions by helicopter, land them directly in combat, and keep them supplied by air.

When the experiment succeeded, the unit was reflagged as the **1st Cavalry Division (Airmobile)**. For the first time in Army history, helicopters were not just support but the centerpiece of operations. Freeman became executive officer of **Company A, 229th Assault Helicopter Battalion**, serving under Major Bruce Crandall.

The two men could not have been more different in temperament. Crandall was outspoken, fiery, and quick to argue with higher headquarters. Freeman was quiet, deliberate, and more comfortable listening than talking. But in the cockpit, they made an ideal pair. Crandall pushed, Freeman steadied. Pilots learned they could trust both men, one to fight for them with commanders, the other to stand beside them with calm professionalism.

Ia Drang from Freeman's Cockpit

On **November 14, 1965**, Freeman lifted into Landing Zone X-Ray with the first waves of Hal Moore's 1st Battalion, 7th Cavalry. The clearing was quiet at first, but by midday the enemy struck in force. Mortars and machine guns raked the LZ. Moore's men dug in as casualties mounted.

Crandall and Freeman volunteered to bring supplies in and carry wounded out. Each run meant flying into a gauntlet of fire. The approach was lined with bullets, mortar rounds bracketed the landing zone, and every touchdown risked disaster.

Freeman's Huey followed Crandall's into the zone. Skids brushed the grass as soldiers rushed to unload ammunition and cases of water. Wounded were dragged aboard, blood pooling on the cabin floor. Freeman never raised his voice, never betrayed fear. His crew remembered, "Captain Freeman made you believe you'd make it. He was calm every time, no matter how hot it got."

During one run, after an intense fire, Crandall considered grounding the company to preserve the aircraft. Freeman looked at him and said, "If we don't go, they die." That quiet insistence carried weight. Both men climbed back into their Hueys and went again.

By the end of the day, Freeman had flown more than twenty sorties. His ship bore bullet holes, his crew was exhausted, but dozens of men were alive because he had refused to stop.

The Quiet Hero

Freeman's courage was not flashy. He did not argue with commanders or bark orders over the radio. He kept flying. For the wounded in his cabin, that calm presence mattered as much as the lift itself. One survivor later said, "I looked up and saw him in the cockpit, tall as ever, steady on the controls. I thought, if he's not scared, maybe I'll make it."

Where Crandall embodied fire, Freeman embodied steel. Both were necessary, both unbreakable.

After Ia Drang

Freeman continued to serve after Ia Drang but retired in 1967 as a major. He had flown through history and came home without fanfare. He settled in Idaho, where he worked as a civil engineer. His projects ranged from road construction to dam maintenance. To his neighbors, he was "Mister Ed," the tall man who fished local streams and helped with community events. Many had no idea he was a hero of the Vietnam War.

He raised a family with his wife, Barbara, was proud of his two sons, and led a quiet life. His humility was as consistent as his flying had been. He did not brag, he did not push himself forward, he lived.

Belated Recognition

In 1992, Hal Moore and Joseph Galloway's book *We Were Soldiers Once... and Young* brought Freeman's name back into the spotlight. Survivors of Ia Drang pressed for recognition. They testified that his flights had saved their lives.

In **2001**, more than thirty-five years after the battle, Freeman stood in the East Room of the White House as President **George W. Bush** placed the **Medal of Honor** around his neck. The citation credited him with "conspicuous gallantry and intrepidity" for his repeated flights into Landing Zone X-Ray, saving dozens of soldiers.

Freeman accepted the medal with humility. He thanked his comrades, spoke briefly, and returned home. His neighbors in Idaho said the ceremony did not change him. He remained the same quiet man who preferred fishing to public attention.

Freeman accepting the Medal of Honor from President Bush
White House photo Public Domain

For conspicuous gallantry and intrepidity at the risk of his life above and beyond the call of duty: Captain Ed W. Freeman, United States Army, distinguished himself by numerous acts of conspicuous gallantry and extraordinary intrepidity on 14 November 1965 while serving with Company A, 229th Assault Helicopter Battalion, 1st Cavalry Division (Airmobile). As a flight leader and second in command of a 16-helicopter lift unit, he supported a heavily engaged American Infantry battalion at Landing Zone X-Ray in the Ia Drang Valley, Republic of Vietnam. The infantry unit was almost out of ammunition after taking some of the heaviest casualties of the war, fighting off a relentless attack from a highly motivated, heavily armed enemy force. When the infantry commander closed the helicopter landing zone due to intense direct enemy fire, Captain Freeman risked his own life by flying his unarmed helicopter through a gauntlet of enemy fire time after time, delivering critically needed ammunition, water, and medical supplies to the besieged battalion. His flights had a direct impact on the battle's outcome by providing the engaged units with timely supplies of ammunition critical to their survival, without which they would almost surely have experienced a much greater loss of life. After medical evacuation helicopters refused to fly into the area due to intense enemy fire, Captain Freeman flew 14 separate rescue missions, providing life-saving evacuation of an estimated 30 seriously wounded soldiers -- some of whom would not have survived had he not acted. All flights were made into a small emergency landing zone within 100 to 200 meters of the defensive perimeter where heavily committed units were perilously holding off the attacking elements. Captain Freeman's selfless acts of great valor, extraordinary perseverance, and intrepidity were far above and beyond the call of duty or mission and set a superb example of leadership and courage for all of his peers. Captain Freeman's extraordinary heroism and devotion to duty are in keeping with the highest traditions of military service and reflect great credit upon himself, his unit, and the United States Army.

Final Years and Legacy

Ed Freeman died on **August 20, 2008,** at the age of eighty. He was buried with full military honors at the Idaho State Veterans Cemetery. Hundreds attended his funeral, from family and friends to veterans who owed their lives to his courage.

His legacy endures not only in the Medal of Honor he received but also in the memory of the men he saved. In the story
of Dust Off and airmobile aviation, Freeman stands as the quiet
professional, the man who kept coming back without fuss.

Crandall was the fire, Freeman was the steel. Both carried
forward the creed that Charles Kelly had spoken a year earlier.
"When I have your wounded" was not just a line. It was a
promise, and Ed Freeman kept it, too tall to fly, too brave to
quit.

Notes:

□ **Primary Sources:** After Action Report, 1st Battalion,
7th Cavalry Regiment, November 1965; aviation mission logs,
Company A, 229th Assault Helicopter Battalion, 1st Cavalry
Division.

□ **Official Histories:** Harold G. Moore and Joseph L. Galloway, *We Were Soldiers Once... and Young* (Random House,
1992), for detailed accounts of Freeman's flights at Landing
Zone X-Ray; U.S. Army Center of Military History, *Pleiku
Campaign* monograph.

□ **Published Works:** John M. Carland, *Stemming the Tide:
May 1965 to October 1966* (CMH, 2000); Shelby L. Stanton,
Anatomy of a Division (Presidio, 1987).

□ **Veteran Accounts:** Interviews with Ed Freeman and sur-
viving crewmen, published in the Vietnam Helicopter Pilots
Association oral history archives and regional newspapers following his Medal of Honor award.

□ **Supplemental Sources:** Medal of Honor citation for Capt. Ed W. Freeman, Department of Defense, July 2001.

CHAPTER SEVEN

Jerome Daly: The Relentless Pilot and the Priest

The Mekong Delta was shrouded in haze on **Easter Sunday, March 26, 1967**. Near Vinh Long, three American helicopter crews had been shot down and were stranded in a swampy field. Two battalions of Viet Cong, dug in along the treelines, were closing on the wrecks. Ground reinforcements were hours away. The trapped crews had minutes.

Dustoff pilots weighed the odds. To go into that zone meant flying straight into overlapping fields of fire. Commanders marked it "no-go." Then a calm voice volunteered. **Chief Warrant Officer Jerome R. Daly**, flying with the **121st Assault Helicopter Company**, keyed his radio: *"I'll make the run."*

Over the next hour, Daly flew twelve passes at treetop level, less than a hundred meters from Viet Cong guns. He laid smoke screens, shielded rescue ships, and drew fire onto himself. When he finally landed, his Huey was riddled with more than fifty bullet holes, hydraulics leaking, beyond repair. But the downed crews were alive.

That day earned him the **Distinguished Service Cross**, one of more than **80 decorations** he received during his service in Vietnam. For his crews and the soldiers he rescued, it was just Daly being Daly, relentless, fearless, unshakable. For Daly himself, it was simply what the mission demanded. Years later, wearing a Roman collar instead of flight wings, he would speak of that day with the same humility he brought to his priesthood.

Roots in Faith and Service

Jerome Richard Daly was born in **1931 in Oakland, California**, but grew up in Philadelphia. He was the middle of three brothers, a family shaped by Catholic faith and immigrant grit. His brothers would predecease him, leaving him the one to carry the Daly name forward.

At **Lower Merion High School**, he was known as quiet but dependable. He graduated in 1950 and then attended **St. Joseph's College** in Philadelphia, completing his studies in 1954. His classmates remembered him as studious, polite, and steady — not the type to seek attention, but the one you counted on when things mattered.

Between school years, he worked as a volunteer firefighter. Later, he test-flew aircraft for Bell Helicopter and even sold insurance for a time. Each job was characterized by a single theme: service and reliability. He was drawn to roles where lives depended on him.

Daly enlisted in the Army in 1949, re-enlisted in 1958, and by the mid-1960s, he had established himself in Army aviation. Colleagues called him "a pilot's pilot." He combined technical mastery with a temperament that stayed level under stress.

Relentless in the Cockpit

When Daly arrived in Vietnam, his reputation grew quickly. He logged **more than 2,000 hours across three tours**, often flying in conditions others refused.

He wasn't reckless. He was deliberate. Crews recalled his pre-flight briefings: short, focused, calm. On the radio, his voice never rose. When tracers laced the windshield, he kept his tone steady.

Infantrymen came to know his call sign. Survivors said, "If Daly was inbound, you were going home." He never turned away. If the call came, he went.

One fellow pilot described him as "relentless. He did not calculate whether it was possible. He just asked, 'Are there wounded?' If the answer was yes, he flew."

That relentlessness would be tested on Easter Sunday, 1967.

First Run

The Huey dipped low over the paddies, blades chopping the humid air. From his seat, Chief Warrant Officer Jerome Daly could see the wrecks scattered below, three helicopters sprawled in broken shapes across the mud. Men were waving frantically near the downed ships, trying to flatten themselves against the earth as enemy fire raked the open ground. Two Viet Cong battalions were dug into the tree lines; their machine guns already stitched onto the approach.

"Taking fire, left side!" the crew chief shouted as the first bursts slammed into the fuselage. Tracers tore past the cockpit, leaving red streaks so close that they lit up the glass with flashes of light. The Huey jolted as rounds punched through the aluminum skin.

Daly's hands never wavered on the controls. He held the nose level, calm voice over the intercom: "Steady. We're going in."

He skimmed across the mud, skids almost brushing the flooded paddy. From the door, his medic pitched smoke grenades into the open, canisters hissing as white plumes blossomed between the downed crews and the tree line. The screen thickened, curling upward, a fragile wall between friend and enemy.

"Smoke's down!" the crew chief called. Daly pulled pitch, the Huey clawing skyward as enemy fire followed them out. Behind, the stranded men cheered, knowing the rescue had begun.

Second Run

Daly banked wide to the south, trading altitude for speed as he lined up a new approach. The treeline ahead looked quiet for a heartbeat, then muzzle flashes winked to life, and the air filled with the flat, metallic crack of machine guns. A burst of fire hammered across the right door, punching ragged holes while whistling in the slipstream. "Right side taking hits," the crew chief called, already leaning out to scan the ground for the stranded men. The downed crews were huddled behind a dike, one man waving a panel, the others crouched low as rounds clipped mud from the berm. Daly eased the nose down, hands light on the cyclic, voice level on intercom, "Smoke ready." The medic lobbed two canisters long, white fingers of cover boiling between treeline and wreckage. Tracers pushed higher, searching for the Huey, but Daly held the line until the smoke thickened, then pulled pitch, climbing shallow. In the cabin, the medic slid to a knee beside a wounded man from the first run, pressing a bandage tight as the helicopter clawed away. Below, silhouettes moved under the new veil, and somewhere on the ground a voice shouted with relief that someone had come back.

Third Run

The mortar tubes found their rhythm as Daly rolled in again, dull thuds stepping across the paddies, geysers of mud rising in tight succession along his inbound track. The Huey lurched as a near miss slapped the airframe. The copilot braced a knee under the panel and kept his eyes on the gauges, the hydraulic needle twitching, as the oil temperature crept up. "Green across," he reported, though neither man liked what the instruments hinted. "Steady," Daly said, nudging left to ride a slight crosswind. A wrecked helicopter lay ahead, tail boom snapped, rotor head canted at an odd angle. Two crewmen were prone beside it, one dragging the other by the collar toward the shallow fold of the dike. The medic pitched a smoke grenade short, saw it bounce and roll, then sent another farther out to thicken the screen. The door gun spat a short burst toward a muzzle flash in the brush, only enough to make the gunner duck. Rounds rattled through the lower panels, tracers within the cabin like angry fireflies. Daly kept the skids just above the water, held long enough to paint the field with smoke, then eased out, making a shallow climb, the rotor thumping hard, the ship shivering but obedient.

Fourth Run

They had mapped his line now, the enemy guns walking their fire into the corridor they knew he would use. Daly slid the Huey wide, then brought it in low from an unexpected quarter, following a narrow irrigation ditch that ran diagonally toward the crash site. The tree line flared anew when he appeared where no one expected him. Bullets chewed the water in a zipper, sprays rising and falling across his path. "Left door, watch your arc," the crew chief warned, mindful of friendlies near the wrecks. Daly leveled, voice calm, "Smoke left, smoke right." Two canisters tumbled and bloomed in quick succession, a white wall swelling between jungle and men. A round cracked through the cockpit side window, peppering the copilot with glass. He flinched, then gave a tight nod, "Good," a single word to confirm he was fine. On the mud below, one of the downed pilots, face streaked with grime, raised a thumb and pointed toward the thicker cover. Daly saw the gesture, stayed another two beats that felt like a minute, then lifted away. Behind them, rescue birds edged closer, emboldened by the sudden silence between the guns and the survivors.

Fifth Run

The radio crackled with warnings from the rescue flight, telling him the screen was working, telling him to stay clear, telling him he had done enough. Daly acknowledged, then turned back in. The copilot frowned but said nothing, knowing this pilot's rhythm by now, one careful decision laid atop another. Mortars pulsed again, closer than before, and the Huey rode the concussions like a small boat on a rough river. "Hydraulic droop," the copilot called, watching the needle settle lower. "We will manage," Daly said, and brought the nose down to skim the flooded field. The medic already had two canisters in hand, one with a broken spoon that he held closed with two fingers until it was launched. He popped them in a stagger that left gaps where the wind could not steal the plume. From the door, the crew chief caught sight of movement in the brush, a figure rising to a knee with something long on the shoulder. He fired a short, sharp burst, and the shape dropped away. Daly rode the line until the haze looked deep enough, then lifted and slid out sideways to keep the smoke screen intact. Far below, figures moved again, a chain of men hauling another crewman by his web gear.

Sixth Run

The Huey rattled now with a constant vibration, a tired machine asking for mercy it would not get. The crew chief touched the mast with two fingers as if to soothe it, then leaned back out into the wet slipstream. "More fire from the east," he called. Daly chose the more challenging path, turning to run that edge of the field where the guns were hottest. Tracers stitched a bright seam that climbed toward them, then bent under the rotor wash and snapped past the belly. The medic slid the cabin door fully open, braced a foot against the frame, and sent the first smoke deep into the gap between wreck and trees. A second grenade landed short, rolling to the dyke. That minor impediment helped keep the men safe from the shrapnel. The pressure wave hit the helicopter like a giant fist, blowing out a right-side panel. The crew chief reported, "Right side is blown through," as he looked for hydraulic leaks. Daly held past his normal mark, counting off seconds in his head, then eased the collective up, the Huey lifting heavily, blades biting thick air as he traded risk for the cover his smoke would buy.

Seventh Run

They had reached the point where crews began to argue, a quiet threshold that hung inside their headsets. "Sir, she cannot take another," the copilot said, eyes on the hydraulic gauge, needle low and jittering. Daly's reply was patient, "One more." He swung wider than before, trying to throw off the mortarmen who had learned his timing, then rolled the Huey in toward a notch in the trees that promised a little shadow. The treeline erupted anyway. Bullets ticked across the cowling, the sound like stones thrown at a tin roof. The medic launched smoke high to let it fall farther forward, the canister bouncing and spinning out a thick white braid that climbed and fattened in the damp air. "Hold, hold," the crew chief urged, watching a wounded man crawl from the wreckage to the lee of a dike. Daly held as long as he dared, then lifted with a softness that spared the rotor head a jolt. The Huey shivered, then found a slow climb, nosing across the screen he had just laid. On the ground, a downed gunner rolled to his back and laughed once, an odd bark at the sky, because the same helicopter had returned seven times.

Eighth Run

Cloud shadow slid across the paddies as Daly came again, this time almost parallel to the tree line, a path that forced the gunners to lead him more precisely. The enemy tried anyway, streams of tracer reaching, then falling short. A heavier weapon spoke from deeper in the brush, a chopping cadence that sent thicker sparks off the far side door. "New gun in there," the crew chief said, and the door gun answered with a controlled string that kept heads down without wasting precious rounds. Daly flew close enough to see the texture of bark along the edge of the grove, close enough to feel the air roughen over the trees. The medic put a canister right where the gunners hated it most, a white knot that billowed and spread along the line. Smoke curled into the branches and spilled down in sheets. In the haze, a rescue bird darted in to snatch two men and was gone, the timing a dance made possible by the cover. Daly rose off the deck and slid away, a thin wake across the surface of the flooded field, his jaw set, his hands unhurried.

Ninth Run

The smoke lay heavy now, low over water and mud, rising into the trunks and blurring the enemy's sight lines. Daly used it the way a river pilot uses a bend, easing the Huey into and along the white to keep the tree line blind. Mortars still hunted him, thumps stepping in behind, then ahead, then behind again, but the bursts were guesses now. "Panel, eleven o'clock," the crew chief called, spotting another survivor signaling weakly from a shallow fold. Daly drifted toward it, nose just a little high, skids a hand's breadth from the flooded field. The medic threw long, a canister landing beyond the man and bursting into a fresh wall that hid his crawl. A burst chewed the lower skin near the battery bay, a harsh metallic rattle that made everyone flinch. "Through the belly, no leak," the crew chief reported, quick eyes picking up what mattered, ignoring what did not. Daly stayed until the new smoke blossomed fully, then lifted and angled away, leaving behind a minor miracle, a living corridor where a man could move unseen.

Tenth Run

Warning lights had come on the panel: hydraulic pressure low, the torque needle flickering on rough pulls, and the generator light threatening to come on. "Engine compartment shows a hit," the copilot said, tapping the gauge with a knuckle. Daly drew a more extended breath, then put the nose down. The approach felt slower, the air thicker, the ship more stubborn than before. The tree line answered with everything it had, streams of fire clawing at the Huey as if trying to pull it from the sky. The medic timed his throws to take advantage of the gaps he had identified throughout the morning, placing canisters where the wind would carry their cover in the right direction. The crew chief spotted two men dragging a third, one of them pausing to lift a hand in a gesture that was part salute and part plea. Daly gave them five long seconds more than he wanted to provide, five seconds that tested the rotor and the nerves of everyone aboard, then eased the collective up and felt the helicopter respond like an old horse asked for one more hill. They rose, just enough, and crossed their own smoke into a space where the fire thinned.

Eleventh Run

The Huey shook so hard now that the compass card went blurry, the needle wobbling without conviction. Every loose object in the cabin found a rattle, a tinny chorus that made the ship sound hollow. "Sir," the copilot said, not finishing the sentence that had lived in his mouth since the seventh run. Daly's answer was the same, quiet and final, "Until they are out." He flew a crooked line on purpose, a gentle snake that made the gunners adjust faster than they wished, then flattened the path as he neared the field so his medic could place the smoke, one to the near side, one to the far, pinning a strip of white between tree line and wrecks. Rounds clipped the rotor arc, a sound that made every man hold his breath for an instant. The crew chief counted two silhouettes moving under the new cover, then lost sight of them behind the veil. "They are going," he said, and it sounded like a prayer. Daly carried the Huey straight through the boiling wall and out the far side, where the air felt briefly cooler, then began the slow climb again.

Twelfth Run

Everything in the cockpit told him they were at the limit, the thrum through the pedals, the weight in the controls, the faint burn of hydraulic fluid in the air. Everything outside said go again. Smoke was thinning, wind shifting, rescue ships queuing for one last dash. Daly set the nose for the same small seam he had used before and let the Huey sink into the low white. The tree line erupted with enemy fire, bullets striking across the aircraft in a steady hail. A round punctured the side panel and slapped the copilot's sleeve without breaking skin, a warm intruder that left a black mark and a startled grin.

The medic sent his last smoke canister and watched it tumble end over end, then burst, covering the gap. "Smoke is down," he said, voice flat with relief. Daly kept the Huey steady for a few seconds more, then drew them out. The ship struggled but reacted to Daly's direction. They cleared the field and climbed toward the hospital pad, the crew suddenly aware of every ache in their bodies, every streak of sweat and grime on their arms. Behind them, the last rescue helicopter rose through the veil with the final survivors aboard, the field at last empty of men who could be saved.

Aftermath

The downed crews were alive. Against overwhelming odds, the rescue had been a success.

Daly's citation for the **Distinguished Service Cross** noted that he "repeatedly exposed himself to intense enemy fire to place smoke screens, enabling the rescue of downed helicopter crews."

Moments before Gen. Harold K. Johnson (left), U.S. Army chief of staff, on behalf of the president, pins the Distinguished Service Cross on Chief Warrant Officer Jerome R. Daly. Soc Trang Airfield, Aug. 4, 1967 (WGBH Media Library and Archives/National Archives and Record Administration (NARA))

It was one of **more than 80 decorations** he would receive in Vietnam: Silver Star, three Distinguished Flying Crosses, two Bronze Stars for valor, two Purple Hearts, and dozens of Air Medals. By the time he left Vietnam, he was recognized as the most decorated Army aviator of the war.

His crews remembered less the medals than the man. "He never once raised his voice," one said. "He just kept saying, 'One more.'"

From Cockpit to Collar

Daly retired from the Army in 1982 as a lieutenant colonel. He could have lived quietly on his reputation. Instead, he began another life of service. He entered seminary and was ordained a Catholic priest in 1987 for the Diocese of Arlington, Virginia.

For nearly two decades, he served parishes across northern Virginia. Parishioners remembered his humility. "You'd never know he was a war hero," one said. "He carried himself as a servant, not a soldier."

Veterans found him, too. Men who had flown with him came to his rectory doors. Families of the fallen asked him to preside at funerals. He ministered to the wounded of both body and spirit.

He retired from active ministry in 2004 but continued to serve until his death on **January 14, 2023**.

Legacy

Jerome Daly's life defies categories: soldier, aviator, rescuer, priest. In Vietnam, he was the man who flew through fire again and again, saving comrades when others thought it impossible. In peace, he was the priest who guided communities and comforted veterans haunted by war.

He earned more medals than any other Army aviator of the Vietnam War, yet he rarely spoke of them. What he spoke of was service, to the wounded, to the faithful, to anyone in need.

Charles Kelly had said, "When I have your wounded." Daly lived those words in war and carried them into his ministry. For Dustoff crews, his name meant courage. For his parishioners, it meant faith.

He was, in every sense, relentless.

Notes:

☐ **Primary Sources:** Mission report, 121st Assault Helicopter Company, 13th Combat Aviation Battalion, March 26,

1967; Army aviation logs for IV Corps, Easter Sunday operations.

☐ **Official Histories:** U.S. Army Center of Military History, *Vietnam Studies: Tactical and Material Innovations* (1973); Dorland and Nanney, *Dust Off: Army Aeromedical Evacuation in Vietnam* (1982).

☐ **Published Works:** Obituaries and biographical sketches in *The American Spectator* (2023) and *Catholic Herald* (2023); Shelby L. Stanton, *Vietnam Order of Battle* (1981).

☐ **Veteran Accounts:** Vietnam Helicopter Pilots Association oral histories mentioning Daly's DSC smoke-screen mission near Vinh Long; parishioner recollections of his later priesthood.

☐ **Supplemental Sources:** Distinguished Service Cross citation for Jerome R. Daly, Department of the Army, 1967; Silver Star citation (date unspecified), U.S. Army Awards Branch.

Chapter Eight

Douglas E. Moore Stephen H. Hammond

AND THE DUSTOFF ETHOS

B y 1968, Dustoff had become part of every soldier's vocabulary in Vietnam. To call for Dustoff meant the difference between life and death. To the infantry, it meant somebody was coming, no matter how destructive the fire was, no matter what the weather looked like. The sound of Huey blades drawing closer through smoke or rain was as crucial to morale as the promise of reinforcements.

The mission itself was brutally simple: take unarmed helicopters into battlefields, pick up the wounded, and get them out. In Vietnam, the red cross painted on the doors rarely offered protection. The Viet Cong and North Vietnamese often targeted Dustoff helicopters deliberately, knowing that taking them out crippled American morale. Pilots and crews learned quickly that the red cross was not a shield, but a target. They went anyway.

The Battles They Flew Into

Dustoff was there at every turning point of the war. At **Ia Drang in 1965**, medevac Hueys carried the wounded out of LZ X-Ray and LZ Albany, sometimes hovering under machine gun fire with skids sunk in mud. In **1967, at Dak To**, Dustoff birds clawed their way into the firestorm on Hill 875, lifting paratroopers out of clearings that were under mortar fire the entire time. In early **1968 at Khe Sanh**, Dustoff crews slipped into a besieged perimeter day after day while artillery and rockets fell across the strip. In **1969 at Hamburger Hill**, they flew through monsoon rain, settling Hueys on slopes so steep that one skid was planted in the mud and the other hung in space.

These were the battles that made headlines. But the daily work was even harsher: single platoons ambushed in elephant grass, villages cut off by enemy fire, convoys torn up on jungle roads, and each of those moments meant a new call for Dustoff. For every high-profile fight, hundreds of missions never appeared in the news.

Tet and the Battle of Hué

The Tet Offensive exploded across Vietnam in January 1968. Hué, the old imperial capital, became the focus of one of the war's fiercest battles. For nearly a month, Marines and South Vietnamese soldiers fought house to house through the walled city, clearing courtyards, temples, and alleyways. By the time the city was retaken in late February, the civilian toll was horrific, and the surrounding countryside was still thick with enemy units.

For Dustoff crews, Huế during Tet was one of the most dangerous places in the war. Roads were blocked, medical facilities were overwhelmed, and casualties mounted faster than they could be moved. Dustoff helicopters flew dozens of missions each day, often landing in courtyards, soccer fields, or even streets cleared just enough to fit a Huey. Enemy snipers targeted them from rooftops. Mortars fell into their landing zones. Every approach felt like a gamble.

It was into that environment that Captain **Douglas E. Moore** flew on **March 2, 1968**.

Douglas E. Moore's Distinguished Service Cross Mission

Moore was serving with the **82nd Medical Detachment (Helicopter Ambulance), 67th Medical Group**. On that day, he was ordered to evacuate American soldiers trapped near Huế. The radio reports were grim: multiple wounded, the landing zone under fire, weather closing in.

The First Run.

Moore nosed his Huey down through gray skies toward the clearing. Small arms fire began before he even flared. Mortars thumped around the zone, dirt and smoke rising in plumes. His crew chief braced in the door, waving men forward as Moore held the helicopter steady. Soldiers ran through the mud, dragging their wounded by webbing straps. They heaved the first casualties across the skid, boots and rifles clattering on the floor. The medic dropped to his knees immediately, tearing open bandages, pressing his palms flat against the sucking chest wounds. The gunner fired in short bursts, muzzle flashing as tracers arced into the tree line. With the cabin loaded as far as space allowed, Moore lifted. The Huey bucked as rounds slapped against the fuselage, but he climbed away.

The Second Run.

The easy choice would have been to break off; the aircraft was already hit. Instead, Moore turned back. More men were waiting. Again, he brought the Huey down into the fire. The windshield cracked white as bullets punched through Plexiglas. The fuselage sprouted new holes. Mortars walked closer to the zone. Fragments wounded Moore himself, but he steadied his hands on the controls and held his hover. More casualties were pulled inside, mud streaking across the floor, blood pooling around the medic's knees. The crew chief shouted that they were full, but Moore refused to leave until he was certain no more could be carried. He pulled pitch, skids dragging mud as the Huey clawed into the air a second time.

The Third Run.

Back at altitude, the reports came in: more wounded were still on the ground. Moore's aircraft was damaged, warning lights flickered across the panel, and he was bleeding himself. He turned anyway. A third time, he brought his Huey into the same deadly landing zone. The crew cleared what little space they had left, ammo cans shoved into corners, ration boxes kicked out the door. Men were hauled aboard one by one, laid across rocket boxes, and tied down with straps. The medic crawled between them, his hands slick, shouting for more dressings; he could not hear over the turbines. Enemy fire raked the helicopter, but Moore held steady until the last casualty was loaded. Only then did he pull away for the final time.

When the helicopter returned to base, mechanics counted dozens of holes across the fuselage and tail. It was grounded for days of repairs. Moore was treated for wounds of his own. His **Distinguished Service Cross citation** recorded the facts with military precision: "Despite adverse weather conditions and intense hostile fire, Captain Moore repeatedly flew into a landing zone surrounded by enemy troops to evacuate severely wounded soldiers. His aircraft was heavily damaged, and he was himself wounded, but Captain Moore refused to abandon the mission until all casualties had been removed to safety."

For the soldiers who watched his Huey drop into the kill zone three times, the formal language meant less than the memory. They remembered a pilot who would not leave without them.

A Career in the Cockpit

Moore's career went far beyond that single day. He flew **1,874 combat missions** and evacuated nearly **2,800 patients**. He received **two Distinguished Flying Crosses** for valor on other missions, as well as a **Purple Heart**.

Flying Dustoff in I Corps was a constant trial. The weather was brutal: sudden monsoon downpours, fog that hid ridgelines, ceilings that forced pilots to skim treetops. Missions were flown at night under flares or by the light of tracers and muzzle flashes. Crews rarely knew if the zone they were heading into had been cleared. Moore lived that tempo daily.

Each mission took its toll. Crew chiefs later described scrubbing blood, mud, and hydraulic fluid out of Huey floors with kerosene rags, only to launch again the next day. Pilots like Moore carried exhaustion in their faces but flew anyway. The numbers alone show the truth: 1,874 missions meant almost no day in his tour passed without at least one flight into danger.

Stephen H. Hammond – Silver Star in II Corps

If Moore's mission showed the Dustoff spirit at Huế, First Lieutenant **Stephen H. Hammond** carried the same creed into the Central Highlands. In **April 1967**, Hammond was flying with the **498th Medical Company** when he was called to evacuate casualties from a firefight in II Corps.

The area was a maze of ridgelines and valleys, the weather closing in. Enemy fire swept the landing zone. Hammond's official **Silver Star citation** recorded that he "made repeated approaches into the landing zone, fully aware of the intense hostile fire directed at him, and safely removed the wounded."

The First Approach.

The radio call guided him toward a scar of clearing in the jungle. Even before he flared, enemy guns opened. Tracers laced the sky, punching holes through the airframe. Hammond steadied the Huey, held it long enough for the first group of wounded to be carried forward. The crew chief and medic dragged them aboard, laying men on the floor, strapping them down with rifle slings. The medic went to work instantly, hands already soaked with blood. The door gunner laid suppressive fire as Hammond pulled away, the helicopter rattling from hits.

The Second Approach.

Reports came back that more men remained. Hammond circled and returned. A second time, he flew into the zone. The enemy fire was heavier, now that he was expecting it. Bullets tore through the cabin wall, showering sparks. Still, he held steady. More casualties were loaded, one man nearly collapsing on the skid before the crew chief pulled him inside by his harness. The medic shouted for morphine; he could barely hear over the turbines. With the deck slick and crowded, Hammond lifted again, coaxing the battered Huey back into the sky.

The Third Approach.

By the time he came back for a third approach, the helicopter carried scars all across its skin. Fuel seeped from a nicked line, and the crew knew the risk of fire was real. Hammond pressed on. He flared hard into the clearing once more, holding steady while the last of the wounded were brought forward. The crew cleared space by shoving ammo boxes aside, wedging men wherever there was room. The medic tied a casualty upright against the frame because the floor was already packed. Rounds cracked through the cabin, one grazing the crew chief's helmet. Hammond did not lift until the ground force shouted that all wounded were aboard.

When he finally climbed out for the last time, the Huey carried more patients than it was designed for. The controls felt heavy and sluggish, but Hammond managed to keep the plane airborne until he reached safety. His **Silver Star citation** summarized the mission: "Despite heavy automatic weapons fire and adverse weather, he made repeated approaches into the landing zone and safely removed the wounded. His courage and determination saved many lives."

For the soldiers who had lain bleeding in that clearing, Hammond was Dustoff at its purest: unarmed, under fire, and unwilling to quit until everyone was out.

Other Dustoff Heroes – Behrens and Evans

Moore and Hammond were not exceptions. They were part of a fraternity of Dustoff pilots who flew the same way. Among the names remembered in oral histories are **Charles Behrens** and **Gerald Evans**.

Behrens was known for his coolness under fire, often bringing his Huey into rice paddies under heavy machine gun fire while infantrymen dragged casualties through water up to their chests. Evans had a reputation for flying in weather others considered impossible, edging his helicopter along ridgelines in fog to reach wounded on slopes. Both men flew missions that infantry remembered as lifesaving.

Some secondary accounts credit them with higher awards, but their names do not appear on the official rolls of Distinguished Service Cross recipients. They almost certainly earned other decorations, Dustoff pilots as a group received hundreds of Distinguished Flying Crosses and thousands of Air Medals. Behrens and Evans deserve to be remembered not for what medal they wore, but for what they did: risking their lives unarmed, again and again, to pull wounded men out of killing zones.

Infantry and Medic Voices

The most explicit testimony to Dustoff came from the men they carried. A paratrooper who fought in I Corps recalled that he thought he was finished until he heard the faint thump of blades through rain. "That sound meant we had a chance," he later said. A medic remembered handing wounded up to a Dustoff crew while bullets struck around them: "I thought they'd leave us, but they didn't. They stayed until the last man was loaded."

Crew chiefs told their own stories. One said he once tied a casualty to a doorpost with a rifle sling because there was no room left on the floor. Another remembered booting rocket boxes out the door to clear space. "You figured you could always get more rockets," he explained, "but you might not get another chance to pull that guy in."

Every voice carried the same truth: Dustoff came, no matter what.

Reflection

Douglas E. Moore's Distinguished Service Cross mission at Hué and Stephen Hammond's Silver Star mission in II Corps both stand as clear examples of the Dustoff creed. But they were just two among hundreds. Moore, Hammond, Behrens, Evans, and countless others proved the same truth: the Dustoff motto, "When I have your wounded," was not just words. It was how they flew.

Together, Dustoff pilots evacuated more than **900,000 patients** in Vietnam, on nearly **500,000 sorties**. Moore's personal tally — 1,874 missions, 2,800 lives — was staggering. Hammond's courage in the Central Highlands was no less vital. Their stories stand for a generation of pilots, crew chiefs, and medics who made lifelines out of helicopters.

Citation – Douglas E. Moore

The President of the United States of America, authorized by Act of Congress July 9, 1918 (amended by Act of July 25, 1963), takes pleasure in presenting the Distinguished Service Cross to Captain Douglas E. Moore, United States Army, for extraordinary heroism in connection with military operations involving conflict with an armed hostile force in the Republic of Vietnam, while serving with the 82d Medical Detachment (Helicopter Ambulance), 67th Medical Group, 44th Medical Brigade. Captain Moore distinguished himself by exceptionally valorous actions on 2 March 1968 while piloting a medical evacuation helicopter near Hue, Republic of Vietnam. Despite adverse weather conditions and intense hostile fire, Captain Moore repeatedly flew into a landing zone surrounded by enemy troops to evacuate severely wounded soldiers. His aircraft was heavily damaged, and he was himself wounded, but Captain Moore refused to

abandon the mission until all casualties had been removed to safety. His extraordinary heroism and determination saved numerous lives and are in keeping with the highest traditions of the military service and reflect great credit upon himself, his unit, and the United States Army.

Citation – Stephen H. Hammond

The President of the United States of America takes pleasure in presenting the SILVER STAR to First Lieutenant Stephen H. Hammond, United States Army, for gallantry in action in the Republic of Vietnam on 21 April 1967. Lieutenant Hammond distinguished himself by exceptionally valorous actions while piloting a helicopter ambulance on an urgent mission to evacuate casualties from a hotly contested area. Despite heavy automatic weapons fire and adverse weather, he made repeated approaches into the landing zone and safely removed the wounded. His courage and determination saved many lives and reflect great credit upon himself, his unit, and the United States Army.

Chapter Notes – Chapter 8

- **Primary Sources:** Distinguished Service Cross citation, Captain Douglas E. Moore, U.S. Army, 82d Medical Detachment, 2 March 1968. Silver Star citation, First Lieutenant Stephen H. Hammond, U.S. Army, 498th Medical Company, 21 April 1967.

- **Official Histories:** U.S. Army Center of Military

History, *Vietnam Studies: Dust Off – Army Aeromedical Evacuation in Vietnam* (1982).

• **Published Works:** Peter Dorland and James Nanney, *Dust Off* (CMH, 1982); Shelby L. Stanton, *Vietnam Order of Battle* (1981).

• **Veteran Accounts:** Vietnam Helicopter Pilots Association oral histories referencing Moore, Hammond, Behrens, and Evans.

Chapter Nine

Michael J. Novosel:

The Reluctant Warrior, Father and Son

In October 1969, a 47-year-old pilot guided his UH-1 Huey low over the paddies of the Mekong Delta. The air was filled with the metallic snap of machine-gun fire, the sharp crack of mortars, and the steady beat of Huey blades. South Vietnamese soldiers lay sprawled in the mud, some waving weakly, others too broken to move. Most pilots would have turned away. **Chief Warrant Officer Michael J. Novosel Sr.** pressed in.

He flared into the kill zone again and again, fifteen separate landings that day. His ship came back with Plexiglas shattered, hydraulics leaking, and the tail boom riddled with holes. By dusk, he had pulled twenty-nine men from certain death. For most pilots, such a day would stand alone as the defining story of a career. For Novosel, it was one of thousands.

By the time he left Vietnam, he had rescued **5,589 people**, Americans, South Vietnamese, and civilians caught between. To his crew, he was "Pappy," the calm voice that never faltered. To infantrymen in the mud, he was salvation. In Dustoff history, he became the elder statesman, the man who proved courage had no age.

Steel Town Beginnings

Michael Joseph Novosel was born on **September 3, 1922**, in Etna, Pennsylvania, a mill town on the Allegheny River. His parents, Yugoslav immigrants, carried the discipline of the old country into the new. His father worked long shifts at the mill; his mother kept the home with firm faith.

The Great Depression left little room for idleness. Mike hauled newspapers, patched bicycles, and brought home every nickel. But his fascination was always with the sky. He built model airplanes from scraps of wood and followed aviation stories as if they were scripture. When barnstormers flew into Allegheny County, he was there at the fence line, watching every move.

When Pearl Harbor was attacked, he was nineteen. Within weeks, he enlisted in the **Army Air Forces**, determined to become a pilot.

World War II: The Bomber Years

By 1943, Novosel was flying the **B-29 Superfortress**, the largest and most advanced bomber of its time. Training was relentless: mastering four engines, high-altitude navigation, and coordinating an eleven-man crew. The bomber was a fortress in the sky, but its survival depended on discipline and teamwork.

In 1944, he deployed to Tinian in the Marianas. From there, he flew raids over Japan.

One mission to Nagoya stands out. The crew crowded into the briefing room before dawn, the map stretched across the wall, coffee bitter in paper cups. The target photos passed from hand to hand showed sprawling industrial plants. Engines roared as the big bombers lifted off coral runways, climbing heavy with fuel and ordnance.

Hours of ocean passed below them before the coastline of Japan appeared. The intercom crackled: "Flak ahead." Black bursts shook the plane, jagged metal ripping the air. Enemy fighters streaked through the formation, tracers arching. Novosel kept his bomber steady. "Bomb bay doors open," the bombardier called. The payload fell, the sky lit with explosions, and they turned for home.

One engine sputtered on the return. The bomber rattled and dropped altitude, but they coaxed her across hundreds of miles of sea. At Tinian, the crew stepped off exhausted, streaked with sweat, smoke, and relief.

At twenty-two, Novosel was already a veteran of missions that burned cities. He came home decorated and convinced flying was what he was born to do.

Korea and Frustration

He stayed in the Air Force. When the Korean War erupted in 1950, Novosel flew transport and support missions. They lacked the glory of bomber raids but carried their own dangers: brutal weather, primitive strips, enemy fire.

By the mid-1950s, the Air Force shrank. Promotion slowed. Novosel, then a major, was forced into retirement in 1955. At thirty-three, with thousands of flying hours, he was told his career was over. He flew civilian jobs, stayed in the reserves, and raised his family. Outwardly, he had moved on. Inwardly, he was restless.

Back in Uniform

As the situation in Vietnam escalated, he tried to return. The Air Force turned him down, too old. The Army, desperate for helicopter pilots, said yes, but only as a **warrant officer**. That meant dropping from major to technical officer, starting over.

Most men would have balked. Novosel accepted. "I just wanted to fly," he said later. At **42 years old**, he retrained on the UH-1 Huey. His classmates were half his age. They called him "Pappy." Then they saw him fly, and the nickname became a badge of respect.

Everyday Dustoff

By 1964, Novosel was in Vietnam with the **82nd Medical Detachment**. Dustoff was unlike anything in the Air Force. No long briefings. No formation flights. Just a siren and a call: men bleeding, minutes left.

One night, a mission took him to a hamlet lit only by a lantern. He flared into a rice paddy under mortar fire. His

medic darted between huts, carrying villagers and ARVN soldiers alike. "Pappy never asked who they were," his crew chief
said. "He just held the bird steady."

Another day, he lifted civilians beside soldiers: mothers
clutching babies, farmers still in sandals, ARVN with bloody
bandages. The cabin was in chaos. "Mercy didn't wear a uniform," Novosel said.

Once, his Huey came back with more than fifty bullet holes.
Hydraulics sprayed across the deck. Mechanics patched it with
tape and clamps, wiped blood from the floor, and sent him out
again. He shrugged and flew.

These missions rarely made the news. But they built his legend. Relentless, unshakable, calm.

Rescuing His Son's Crew

By late 1968, Dustoff had become a family affair in the
Novosel household. While Michael Sr. was in his forties and
already on his second tour, his son, **Michael J. Novosel Jr.**,
had followed him into Army aviation. Father and son flew in
different detachments, but often in the same sectors of the
Mekong Delta. Crews sometimes joked that the Delta was the
only combat zone in history where you could run into both
generations of a family in the same sky.

On one humid afternoon, Mike Jr. lifted off on what seemed
like a routine mission, an urgent call from an ARVN unit under
fire in the paddies. The weather was typical for the Delta: low
clouds, oppressive humidity, haze that turned everything gray.
His Huey crew, copilot, medic, and crew chief, moved with

the smoothness of men who had done this hundreds of times. They had no idea that in less than an hour, their mission would become part of the Dustoff legend.

As Jr.'s Huey approached the pickup zone, enemy fire erupted from tree lines. The aircraft rocked under the impact. A burst of heavy machine-gun fire tore into the engine compartment. Warning lights flashed across the panel, gauges dropping. The tail rotor shuddered, then failed. The Huey lost lift, nose pitching down. Jr. fought the controls, trying to cushion the fall. The helicopter slammed into a flooded rice paddy, skids collapsing, water spraying across the cabin.

The crew scrambled out, rifles in hand, dragging the wounded they had managed to pick up. The paddy offered no cover. Bullets whipped across the water, snapping mud into the air. "Get down!" the crew chief shouted, crawling to the dike line. They huddled in the muck, firing back, but they knew they wouldn't last long without rescue.

The Mayday call went out: Dustoff down, crew in the open, taking fire. The transmission crackled across the net.

Miles away, **Michael Sr.** heard it. The call sign belonged to his son. In the cockpit of his own Huey, the intercom went quiet. "That's my boy," he said softly. The crew chief remembered the look on his face: a tight jaw and eyes locked ahead. "We all knew right then," the crew chief recalled. "We were going in after his son."

Sr. nosed his Huey toward the coordinates. The crew tightened their straps and checked their weapons. They knew the landing would be hot. As they neared the site, smoke marked the crash. Enemy tracers arced upward, hunting them.

On the ground, Jr. and his crew heard the familiar whump-whump of rotor blades. At first, it was just hope, then disbelief. Through the haze, a Huey descended. As it flared, they saw the pilot. "It's Dad," Jr. muttered, stunned. The sight was surreal: his own father, gray-haired, lined with years, flying into a firefight to pull him out.

Sr. planted the skids in the paddy, bullets ricocheting off the water. His crew chief leaned out, firing bursts to keep enemy heads down. The medic waved frantically, urging Jr.'s crew to move. They sprinted through knee-deep water, dragging their wounded. The crewmen grabbed them by their web gear, hauling them inside.

For Sr., there was no time for words. He held the controls steady, eyes fixed on the gauges, listening to his crew shout "Clear! Clear!" When the last man was aboard, the crew chief slammed the bulkhead. "Up!"

Sr. pulled up on the collective. The Huey staggered, bullets rattling through the tail. For a few seconds, it seemed the bird wouldn't climb. Then it clawed skyward, skimming the dikes, engine straining. Behind them, the crash site disappeared into smoke.

The crew was soaked, muddy, and trembling with adrenaline. Jr. sat against the cabin wall, chest heaving, staring at his father's profile in the cockpit. The surreal truth sank in: his dad had just saved his life.

That night, father and son spoke briefly. Neither man was given to flowery words. "You okay?" Sr. asked. "I'm good," Jr. answered. There was a long pause. Then Sr. said, "That's Dustoff."

It was the only time in the Vietnam War that a father plucked his own son from combat under fire. For the Novosels, it was more than family, it was the creed. If someone was down, you went to help them. Age, rank, relation, none of it mattered. You went.

October 2, 1969 – The Medal of Honor Mission

The Mekong Delta lay under gray skies that morning, the paddies glistening with rain and mud. A South Vietnamese company had been ambushed near Kien Tuong Province, caught in open fields and hammered by automatic weapons and mortars. By the time the word reached the Dustoff detachment, dozens of men were down. The survivors were pinned, out of ammunition, and watching their friends bleed out.

Chief Warrant Officer **Michael J. Novosel Sr.**, already a veteran of thousands of missions, launched his Huey. He was forty-seven years old, the oldest pilot in the detachment, and his crew, half his age, called him "Pappy." They had seen him fly into bad places before. That day would surpass them all.

First Run

From the cockpit, the paddies looked deceptively calm. As soon as the Huey nosed down, fire opened from the dikes, machine-gun bursts, tracers arcing red through the haze. "Taking fire!" the crew chief yelled over the intercom.

Novosel held steady. The skids splashed into the mud. His medic jumped out, dragging the first casualties aboard. Men with shattered legs, chest wounds, blood soaking through fatigues. They packed them on the floor. "We're full!" the crew chief called. Novosel pulled the pitch, Huey groaning as it lifted into the teeth of fire.

Second Run

He circled back. Enemy fire was heavier this time, zeroed in on the approach. Plexiglas cracked, chips of plastic spraying the cockpit. One round tore into the tail boom with a hollow thump. "She's still flying," Novosel muttered, easing the Huey into the same spot.

More wounded scrambled aboard, some dragging themselves, others carried by comrades. The medic pressed bandages against sucking chest wounds, IV lines swaying as the helicopter shuddered. Novosel lifted again, nursing the controls, the Huey shaking but holding together.

Third and Fourth Runs

By the third sortie, the enemy had the range. Mortar rounds thumped, throwing mud into the air around the LZ. One burst rocked the Huey sideways on approach. "Easy, Pappy!" the copilot shouted. Novosel flared hard, steadied, and planted the skids.

The crew chief and medic worked frantically, dragging soldiers in by web gear, stacking them shoulder to shoulder. "We can take two more!" the medic yelled. Novosel gave them the seconds. Then they were airborne, tail rotor chopping spray as another mortar crashed behind them.

The fourth run nearly ended it. A volley of fire ripped across the fuselage, bullets chewing through the floor. Hydraulic fluid sprayed in the cabin, slicking the deck red with oil and blood. The controls went heavy in his hands. "Hydraulics are shot!" the copilot warned. "We'll manage," Novosel answered. He coaxed the bird out, climbing away with another load.

Fifth through Seventh Runs

Crews on the ground were stunned that he returned at all. They had assumed one pickup, maybe two, before the Huey was too damaged. But minutes later, the sound of rotors came again.

On the fifth run, Novosel angled in from a different direction, trying to throw off the gunners. Still, rounds snapped past the canopy. Soldiers waved frantically from shell holes. His medic leapt out again, pulling men aboard one by one. They were running out of litter straps. The wounded were lashed to the cabin bulkhead with belts and rifle slings.

By the seventh run, the Huey rattled with vibration. A round had nicked the rotor mast; causing a severe vibration. "We thought the blades would fly off," the crew chief said later. But Novosel held the hover, eyes locked on the gauges, knuckles white on the controls. Another half-dozen men were dragged aboard before he eased out of the fire again.

Eighth to Tenth Runs

Each circuit back to the hospital pad, mechanics and medics swarmed the ship, tape over bullet holes, bandages over men.

The crew barely had time to wipe blood off the floor before the skids lifted again.

The eighth run was chaos. The enemy concentrated fire on the LZ, determined to kill the Huey. Bullets punched through the cabin, shattering glass, spraying fragments. One round tore into the instrument panel. "We're losing systems!" the copilot warned. "We've still got lift," Novosel said, voice calm. They lifted out again, cabin jammed with men groaning in pain.

The tenth run felt impossible. Smoke and dust obscured the approach. Soldiers crawled through mud toward the ship, some too weak to rise. Novosel flared, hovering inches off the ground while his medic pulled them in by their gear straps. "We were hauling them in by the handful," the medic said later. The Huey staggered into the air, overloaded, straining to climb above the trees.

Eleventh to Thirteenth Runs

By now, the Huey was a sieve. More than fifty holes scarred the fuselage. The crew was exhausted, soaked in sweat and blood. The aircraft trembled in flight, every approach a gamble.

On the eleventh run, a mortar landed so close that mud splattered the cockpit glass. The shock nearly knocked them off balance. "Hold her!" the crew chief shouted. Novosel steadied, planted skids, and they loaded more.

The twelfth run, the crew finally argued. "Sir, she won't take another," the copilot said. "One more," Novosel replied evenly. He brought them in low again, Huey shuddering.

On the thirteenth run, the vibration was so violent that the compass spun uselessly. The Huey felt like it was about to shake apart. Yet they lifted three more men, one missing both legs, another unconscious, face ashen. "Go!" the medic yelled, pounding the bulkhead. Novosel pulled them out again.

Fourteenth and Fifteenth Runs

By the fourteenth run, fuel was low, controls were stiff, and systems were failing. The crew begged him to stop. Novosel shook his head. "There are still men down there."

He came in one last time, the fifteenth approach. Enemy fire erupted, louder than ever. The Huey shook, tail boom rattling, cockpit gauges flickering. His medic scrambled into the mud, dragging the last survivors aboard. The crew chief hauled them in, shouting, "That's all of them!"

Novosel lifted, muscles straining against stiff controls, the Huey lurching upward like a wounded bird. They staggered over the dikes, nearly clipping trees. Somehow, they made it out.

After the War

Novosel retired in 1976. He had flown **2,500 combat missions**, rescued **5,589 people**, and earned the Medal of Honor, Distinguished Service Cross, Distinguished Flying Crosses, Purple Hearts, and more than fifty Air Medals.

Novosel wearing the Medal of Honor earned in Vietnam Publis Domain

He became a test pilot, training younger aviators. At reunions, crews sought him out. They called him "the grandfather of Dustoff." He always deflected praise to medics and crew chiefs.

He was proudest of serving alongside his son. Very few families endured as much or maintained such a strong connection during times of conflict.

Final Years and Legacy

Michael J. Novosel Sr. passed away on **April 2, 2006**, at the age of eighty-three. He was buried at Arlington National Cemetery. His coffin rolled past rows of white stones. His son saluted at the grave.

His story is singular: a WWII bomber pilot turned helicopter rescuer, a father who saved his own son, a man old enough to be a grandfather flying into kill zones until the last man was out.

For infantrymen, his Huey was salvation. For Dustoff, he was proof that courage has no age.

"When I have your wounded" was the creed of Charles Kelly. Michael Novosel kept it **5,589 times**.

Notes:

- **Primary Sources:** Mission logs, 82nd Medical Detachment (Helicopter Ambulance), 1968–1969; operational reports from the 13th Combat Aviation Battalion, IV Corps Tactical Zone.

- **Official Histories:** Peter Dorland and James Nanney, *Dust Off: Army Aeromedical Evacuation in Vietnam* (Center of Military History, 1982); U.S. Army Aviation Digest articles, 1969–1970, covering Dustoff operations in the Delta.

- **Published Works:** Michael J. Novosel with Stephen Coonts, *Dustoff: The Memoir of an Army Aviator* (Presidio, 1999); Shelby L. Stanton, *Vietnam Order of Battle* (Galahad Books, 1981).

- **Veteran Accounts:** Vietnam Helicopter Pilots Association oral histories, including recollections of Novosel's Medal of Honor mission on 2 October 1969; interviews with Michael J. Novosel Jr. describing father-and-son service in Vietnam.

- **Supplemental Sources:** Medal of Honor citation for Chief Warrant Officer Michael J. Novosel Sr., Department of Defense, 197

CHAPTER TEN

Patrick Brady: The Mission Never Stops

On the morning of **January 6, 1968**, Captain Patrick Henry Brady strapped into the right seat of his UH-1 Huey near Chu Lai. Outside, the weather was a wall of fog and rain, thick enough to ground most aircraft. Inside, the radio carried urgent calls. American units were pinned down, surrounded, and bleeding. Medevac was needed. Other pilots had turned back. Brady volunteered.

Over the next twenty-four hours, he and his crew would fly **six separate missions** into zones swept by fire, through fog so thick that treetops brushed his skids, into minefields and swamps. By day's end, he had evacuated more than **fifty-one wounded soldiers**, many from places no other aircraft would go. His Medal of Honor citation would later say he "unhesitatingly risked his life to save others." His men remembered it more simply: "He never quit. If you were hurt and called Dustoff, Pat Brady was coming."

An Irish American Beginning

Patrick Henry Brady was born on **October 1, 1936**, in the small town of **Philip, South Dakota**, to a Catholic family of Irish descent, one of three children. His father, a World War I veteran, worked in construction jobs and sometimes struggled to find steady employment. His mother anchored the household with faith and discipline.

When Pat was still young, the family relocated to Seattle, Washington, in search of new opportunities. There he grew into a wiry, competitive teenager who thrived on challenges. At **O'Dea High School**, a Catholic boys' school, Brady played football and basketball. Coaches and teammates recalled him as quick to laugh off bruises, stubborn in defense, and fiercely loyal to his team. The values of faith, loyalty, and discipline instilled in him at home and at school became his guiding principles.

In 1954, Brady enrolled at **Seattle University**, a Jesuit institution. He studied political science, but the real turning point came in the Reserve Officers' Training Corps program. At first, he saw ROTC as a method to pay for school, nothing more. By the time he graduated in 1959, he had found his calling. Commissioned as a second lieutenant, he decided to make the Army his career.

Into the Army and Toward the Sky

Brady's first assignment was in field artillery. He learned gunnery, tactics, and the discipline of soldiering, but it was aviation that caught his eye. The Army was rapidly expanding its rotary-wing fleet, and helicopter pilots were in demand. Brady applied for training and was accepted.

At **Fort Rucker, Alabama**, he earned his wings in 1963. He fell in love with flying, mastering the Huey, a machine that seemed an extension of the pilot's will. Brady later said the Huey was "the most perfect combat machine ever built for what we needed in Vietnam."

In 1964, he was sent to Vietnam for his first tour. There, he came under the command of **Major Charles L. Kelly**, the fiery leader of the 57th Medical Detachment who had set the Dustoff creed. Brady flew alongside him, learning what it meant to risk everything for the wounded. When Kelly was killed in July 1964, his dying words, *"When I have your wounded,"* became Brady's personal mantra. He carried them into every mission that followed.

Back to Vietnam

After his first tour, Brady returned home, then volunteered to go back. By late 1967, he was a captain assigned to the **54th Medical Detachment**, based at Chu Lai in I Corps, the northernmost sector of South Vietnam. He was thirty-one years old, experienced, and already a seasoned Dustoff pilot.

I Corps was the hardest assignment of all. The terrain was brutal, from coastal swamps to steep mountains. The weather was worse, with sudden fog banks, torrential rain, and winds that shifted without warning. The enemy was aggressive, often waiting for helicopters to descend before opening up with machine guns and mortars.

Brady developed techniques that became his signature. In fog, he flew low, using treetops, rivers, and even the smell of smoke to guide him. When fire swept the landing zone, he varied his approach, coming in from unexpected directions. He believed every call had to be answered. His crews trusted him completely.

January 6, 1968

The Americal Division was in heavy contact near Chu Lai. The weather was so poor that flights had been canceled, the sky sealed in fog and drizzle. Brady heard the calls for help and launched anyway. Over the course of that day, he would fly six extraordinary missions, each one different, each one dangerous.

First Mission: Into the Fog

The first call came from a patrol cut off in a valley where fog blanketed everything. Other helicopters had turned back, unable to find the unit or see the ground. Brady descended into the soup, flying low enough that treetops brushed the skids. He used the sound of gunfire, the smell of smoke, and the radio voices of the trapped soldiers to find his way.

At last, a clearing opened just wide enough for a landing. His crew worked almost blind, loading men whose uniforms were soaked with blood. One medic recalled, "We could not see twenty feet, but we could hear them. Captain Brady held her steady until we had every man we could fit." He lifted off into the mist, trusting his instincts more than his instruments.

Second Mission: Surrounded by Fire

Word came next from a company encircled by enemy troops. The landing zone was ringed with automatic weapons. Brady nosed the Huey into the hot zone, tracers flicking across the windscreen. His crew chief shouted over the intercom as bullets tore into the fuselage.

Still, Brady held the ship steady as his medic dragged the first casualties aboard. Wounded soldiers stumbled into the cabin, collapsing on the floor as rounds punched through the tail boom. With more than a dozen aboard, Brady pulled pitch, coaxing the wounded bird back into the sky. The hydraulic fluid leaked, but he managed to nurse it home.

His crew counted the holes later. They had no time to patch them before he lifted again.

Third Mission: The Minefield

The most perilous call of the day came from men trapped in a minefield. Several had gone down, and the rest dared not move for fear of triggering explosions. Brady knew no other pilot would risk it, so he went.

He eased his Huey into the clearing, skids settling among the craters. His crew chief and medic crawled out carefully, pulling casualties aboard one by one. "We were on our bellies in the dirt," the crew chief remembered. "The captain just kept saying, steady, steady, we're not leaving anyone."

The aircraft took hits, vibrations shaking the frame, but Brady lifted out with the first load, then returned a second time until every casualty was clear. The risk of setting off a mine was constant, but the men were saved.

Fourth Mission: Hovering in the Swamp

Another desperate call came from a patrol pinned down in swampy ground. There was no place to land. Brady brought his Huey to a hover, skids dipping into the water. He held the helicopter steady while his crew waded waist-deep, hauling wounded aboard.

Rounds splashed around them, some striking the water near the skids. The rotor wash churned mud and spray, soaking everyone. One soldier later recalled, "I remember looking up at that bird sitting in the swamp, thinking there is no way we are getting out alive. Then they pulled me in, and Captain Brady lifted us straight up through fire."

Fifth Mission: Through the Firestorm

By afternoon, Brady was called again to a landing zone already bracketed by mortars. The company was under withering fire, their wounded piling up. Brady brought his ship in low, bullets punching through the floor. His medic dragged men aboard, one after another, until the Huey carried far more than it was designed to lift.

Hydraulic lines were severed, fluid sprayed across the cabin, and the controls stiffened. Brady coaxed the Huey out of the clearing, nursing it over the treetops. When he set down at the aid station, his crew chief shook his head at the holes. "She should not have flown," he muttered.

Minutes later, Brady lifted off again.

Sixth Mission: The Last Run

As daylight faded, a final call came from a unit still in contact. Brady had already flown for hours, his bird scarred from earlier runs, fuel low. But he went.

He slipped into the zone in the half-light, muzzle flashes flickering at the edges of the clearing. His crew loaded the last of the wounded, and Brady pulled the pitch. The Huey staggered upward, shuddering, then clawed free. He delivered the men to safety, bringing the day's total to fifty-one soldiers evacuated.

The Measure of That Day

Brady's Medal of Honor citation reduced the events to a paragraph of formal words. His crew and the men he carried remembered more. They remembered the fog that hid treetops, the minefield that could have blown them all apart, the swamp that threatened to swallow the Huey, the holes in the floor and tail boom, and the steady voice of their pilot.

One infantryman wrote years later, "I owe my life to Dustoff. Captain Brady came when nobody else would. He came again and again until we were all out. That is the only reason I am alive to write these words."

For Brady, the day was not about heroics. It was about keeping the creed alive. "It was never about me," he said later. "It was about the guys on the ground. They called. We came."

Other Missions and Relentless Service

The January 6 missions were only one day in a long tour. Brady flew **over 2,000 combat missions** in Vietnam. He evacuated more than **5,000 wounded soldiers**. His Hueys were shot full of holes countless times. He was wounded himself and survived several crashes. Yet he kept flying.

He developed techniques that became Dustoff standards. He used fog and smoke to conceal his approach. He varied altitudes and angles to confuse enemy gunners. He was known for flying so low that his skids brushed treetops. His crews said he could make the Huey do things no manual ever taught.

Brady also believed in leading from the front. As a commander, he flew as many missions as his men. He never asked them to go anywhere he would not. That example inspired trust and loyalty.

Life After Vietnam

Brady remained in the Army, rising through the ranks. He eventually became a **major general**, commanding the 6th Army at the Presidio of San Francisco. But no matter how high he rose, he never forgot Dustoff. He carried its story into every assignment.

After retiring, he became an advocate for veterans and a public speaker. He wrote *Dead Men Flying*, a book about Dustoff and his experiences. He spoke often about leadership, courage, and faith. His message was simple: service is about others, not yourself.

Patrick Brady's legacy is twofold. On one hand, he is remembered as a Medal of Honor recipient who saved dozens of lives in a single day. On the other hand, he is remembered as the embodiment of Dustoff's creed over an entire career.

For the infantrymen of I Corps, Dustoff meant hope. For the pilots who followed, Brady was the standard. He carried forward Charles Kelly's words and repeatedly proved them.

"When I have your wounded" was not just a phrase. For Brady, it was a mission that never stopped.

Citation

For conspicuous gallantry and intrepidity in action at the risk of his life above and beyond the call of duty, Maj. Brady distinguished himself while serving in the Republic of Vietnam commanding a UH-1H ambulance helicopter, volunteered to rescue wounded men from a site in enemy-held territory which was reported to be heavily defended and to be blanketed by fog. To reach the site, he descended through heavy fog and smoke and hovered slowly along a valley trail, turning his ship sideward to blow away the fog with the backwash from his rotor blades. Despite the unchallenged, close-range enemy fire, he found the dangerously small site, where he successfully landed and evacuated two badly wounded South Vietnamese soldiers. He was then called to another area completely covered by dense fog where American casualties lay only 50 meters from the enemy. Two aircraft had previously been shot down and others had made unsuccessful attempts to reach this site earlier in the day. With unmatched skill and extraordinary courage, Maj. Brady made four flights to this embattled landing zone and successfully rescued all of the wounded. On his third mission of the day, Maj. Brady once again landed at a site surrounded by the enemy. The friendly ground force, pinned down by enemy fire, had been unable to reach and secure the landing zone. Although his aircraft had been badly damaged and his controls partially shot away during his initial entry into this area, he returned minutes later and rescued the remaining injured. Shortly thereafter obtaining a replacement aircraft, Maj. Brady was requested to land in an enemy mine field where a platoon of American soldiers was trapped. A mine detonated near his helicopter, wounding two crewmembers and damaging his ship. In spite of this, he managed to fly six severely injured patients to medical aid. Throughout that day Maj. Brady utilized three helicopters to evacuate a total of 51 seriously wounded men, many of whom would have perished without prompt medical treatment. Maj. Brady's bravery was in the highest traditions of the military service and reflects great credit upon himself and the U.S. Army.

Major General Patrick Brady Public Domain

Notes:

- **Primary Sources:** After Action Reports, 54th Medical Detachment (Helicopter Ambulance), January 1968; Americal Division operational logs, Chu Lai sector.

- **Official Histories:** Peter Dorland and James Nanney, *Dust Off: Army Aeromedical Evacuation in Vietnam* (U.S. Army Center of Military History, 1982); U.S. Army Center of Military History monographs on the Tet Offensive, 1968.

- **Published Works:** Patrick H. Brady with William P. Crotty, *Dead Men Flying* (WND Books, 2012); Shelby L. Stanton, *Vietnam Order of Battle* (Galahad Books, 1981).

- **Veteran Accounts:** Oral histories from Dustoff pilots and crews collected by the Vietnam Helicopter Pilots Association, including testimonies about Brady's repeated sorties near Chu Lai, January 6, 1968.

- **Supplemental Sources:** Medal of Honor citation for Maj. Patrick H. Brady, Department of Defense, 1969.

CHAPTER ELEVEN

Slicks Under Fire: Infantry Extractions in the Highlands

The Central Highlands, 1965 – Plei Me

In October 1965, near the outpost at Plei Me, an American rifle company moved through a long valley where the elephant grass rose above their helmets. The patrol had been quiet, too quiet, the kind of silence that made veterans uneasy. The crack of rifles from the tree line ended it, and almost at once, mortars began to fall in a steady beat. Explosions tore into the ground, sending smoke rising in geysers. Soldiers went down in the first moments, some killed outright, others screaming for help. The company hugged the ground, pinned flat, with casualties scattered across the field.

The radioman crawled deeper into the grass, handset pressed tight to his mouth. He called for Dustoff, but the answer came back hard: every medevac bird was already tied up, too far to reach them. For the men bleeding out in the open, it sounded like the end. Then another voice cut across the net, calm and matter of fact. **Company A, 229th Assault Helicopter Battalion**, was on the way. Slicks would come in for them.

The First Huey – Capt. Tom Rickman

Capt. Tom Rickman, an experienced aircraft commander in A/229, led the flight. His crew that day included a copilot, an unnamed medic who had volunteered to ride along, and door gunner **SP4 Davis**. Rickman dropped his Huey low over the grass, the sound of the blades thumping through the valley like a drum. Almost instantly, the tree line flashed with fire, tracers zipping across their nose. Bullets smacked into the tail boom, the vibration kicking through the pedals into his boots.

SP4 Davis leaned into the doorway, firing quick bursts. He was not trying to win the fight, just to make the enemy duck long enough to buy a few seconds. Rickman flared and forced the skids into the mud. Infantrymen ran from the grass, dragging two wounded between them. One's leg was mangled, and the other held his chest with both hands. Davis and the medic pulled them inside, voices raised above the roar of the engine. "That's it, full up," Davis shouted, banging the bulkhead. Rickman lifted the machine carefully, coaxing it into the air, its blades chewing through the smoke. Fire chased them as they clawed over the trees.

The Second Huey – WO1 Bob Taft and WO1 Jack Swain

Behind him came **WO1 Robert "Bob" Taft** with **WO1 Jack Swain** in the copilot's seat. Taft had the controls, Swain worked radios and navigation. As their Huey broke from the trees, the enemy was ready. A burst smashed into the nose, Plexiglas shattering in a spray. Swain jerked sideways, struck by fragments. Taft stole a glance at him, saw blood on his visor, then locked his eyes forward and kept flying.

He slammed the skids into the clearing. Infantry swarmed forward with another man stretched across a poncho. They slipped in the mud, boots skidding, then shoved him aboard. Two others tumbled in after him, breathing hard, helmets tilted sideways. The crew chief hooked a harness under one man's shoulders and dragged him clear of the doorway. The medic crouched over the casualty, blood already pooling across the deck plates. Outside, fire still poured from the tree line. Taft raised the collective, his Huey groaning as if ready to give up. Swain sagged in his seat, but the ship lifted anyway, dragging itself over the ridge while tracers snapped beneath the tail.

The Third Huey – Rickman Back In

Rickman turned his bird again, ignoring the holes already punched through her. Mortars thumped steadily, one burst so close it rocked the ship sideways. He steadied her with small, gentle movements, his jaw clenched. Sparks exploded as rounds ripped through the panels under the crew seats. The crew chief leaned out, firing to cover the soldiers as they crawled from the grass.

Two came dragging another man between them, stumbling as they pulled him forward. They shoved him into the cabin, then fell in behind him, mud streaking their faces. The medic pushed them flat and shouted for straps. The cabin filled shoulder to shoulder, rifles clattering underfoot. The copilot yelled that they were overweight, gauges creeping high, but Rickman said nothing. He inched the skids through the grass until the blades dug in; then the Huey rose, climbing slowly, branches snapping against its belly before it cleared the trees.

The Fourth Huey – Taft Again

Taft swung wide before cutting back into the clearing. It made little difference. The tree line blinked with fire, bullets hammering the doors, tracers jumping along the cabin wall. Still, he flared, forcing the blades down until the grass flattened beneath him. Men rushed from the haze carrying the last of the wounded. The medic dragged them across the deck, already slick with blood and mud. When there was no more room inside, others clung to the skids, their arms hooked around the struts, their boots slipping. "We are too heavy," the crew chief shouted. Taft glanced once at the gauges, then back at the field. He lifted slowly, the Huey shuddering, soldiers swinging from the skids as fire tore through the clearing. Somehow, the bird clawed skyward, wobbling clear of the ridge with every man still aboard.

On the Ground

For the infantry trapped in the grass, each Huey's arrival was a shock of hope. The sound of rotors grew into a pounding heartbeat, then the dark green shapes appeared, settling into a fire that seemed too thick for anything to survive. Men later said they crawled, limped, or threw themselves toward the helicopters. One survivor remembered it plain: "We thought we were finished. Then the slicks came. Every time one lifted out, we prayed another would follow."

Aftermath

By the end of that fight, four slicks from A/229 had gone into the valley. Each came back with damage. Rickman landed with Swain bleeding beside him. Taft limped home in a bird overloaded with men hanging on the outside. Mechanics counted holes across every airframe, shook their heads at how they had held together.

The company survived because those slick crews ignored the word "no-go." They were not flying red cross ships, yet they carried the same creed in their actions. For the men who lived, the memory of that day was not in the citations or reports, but in the stubborn will of pilots like Rickman and Taft who refused to leave them behind.

Ia Drang Valley, November 1965 – Albany

The fight at Landing Zone Albany came two days after X-Ray. Moore's battalion had been lifted out, and fresh battalions were moving along narrow trails when they walked into the teeth of a North Vietnamese ambush. The enemy had planned it well. Automatic weapons tore through the column from three sides, mortars bracketed the open patches, and within minutes, whole platoons were torn apart. Men dove for termite mounds or shallow ditches, some clawing at the ground with their hands to dig cover. Radios filled with panic. Casualties mounted by the dozen.

Dustoff could not reach them at first. The fire was too heavy, and the landing zones too small. That left the slicks of **Company A, 229th Assault Helicopter Battalion**, the same crews who had carried Moore's men into X-Ray days earlier. This time, they came not with infantry but to pull out the wounded and the living who could still move.

Capt. Bill Beck's Run

Capt. Bill Beck nosed his Huey into a patch of clearing hardly bigger than the rotor span. The moment he broke cover, the tree line came alive, tracers tearing across the valley. Bullets smacked the doors. His crew chief, whose name has been lost to time in the records, leaned out and returned fire in short bursts. Beck kept his hands steady, flaring hard and forcing the skids into the mud.

Soldiers rushed him, some dragging the wounded by their web gear, others half carrying men shot through the legs. They tumbled into the cabin, eyes wide, faces gray. The medic inside dropped to his knees, pressing bandages and shouting for straps. More men clambered on until the cabin was shoulder to shoulder, rifles clattering underfoot. "That is all she will take," the crew chief yelled. Beck nodded once, raised the collective, and pulled the ship free, treetops clawing at the skids as the Huey staggered into the air.

WO1 Bill Garay's Run

Behind him came **WO1 Bill Garay**, flying another slick into Albany. His copilot, also a warrant officer, worked the radios, trying to keep order in the chaos. The approach was straight into sheets of fire. A mortar burst close enough to rock the bird, throwing dirt and smoke across the windshield. Garay pressed on. He set the Huey down hard, rotors chopping through brush.

Men stumbled out of the treeline, dragging bodies, some alive, some not. They shoved them in with no ceremony, only desperation. The crew chief grabbed one man by the straps and yanked him inside, and another clung to the doorway until Garay steadied the ship. The medic was already at work, his knees braced, blood streaking his hands. Garay lifted, over the weight limit, engine groaning. The Huey dragged through the grass, its skids ripping up the earth before it found lift. Fire followed them out, bullets chewing at the tail boom. Somehow, she climbed.

A Rescue of Their Own

One slick came in and found another Huey down, riddled with holes and half on its side. Survivors from both aircraft ran for the new ship, swarming aboard in a rush. The crew chief shouted that they were packed too tight, but the pilot only said, "Hang on." Men were crammed inside; others clung to the skids. When the Huey rose, overloaded, soldiers dangled from the struts, swinging as tracers zipped underneath. It was madness, but they were alive, and they were leaving the kill zone.

Voices from the Ground

Infantrymen who survived Albany never forgot the sound of slicks coming through the smoke. One survivor later said he lay flat in the grass, pressing a hand against his stomach wound, certain he was finished. The battle noise had faded in his head, and what he noticed most was the buzzing of flies and the smell of burned powder. Then, faint at first, he heard the chop of blades. It grew louder until the sound filled the clearing. "I thought I was dreaming," he admitted years afterward, "but when that Huey dropped in I knew somebody had come for us."

Another man remembered hanging from a skid, his boots sliding in the wet mud as the helicopter lifted. He had no strength left, only the fear of letting go. "I did not care if I fell," he said, "I just wanted out of that place."

Aftermath

The fight at Albany was one of the worst single days of the war for the 1st Cavalry Division. More than 150 men were killed, and many more were wounded. The 2nd Battalion of the 7th Cavalry was hit hard. The slicks of A/229 went into that hell repeatedly, pulling out who they could. Pilots like **Bill Beck** and **Bill Garay** flew ships home riddled with holes, mechanics counting dozens of strikes along each airframe.

For the men who lived, the memory of Albany was blood and fire, but also the sight of Hueys dropping into impossible clearings, pilots holding steady, crews dragging them aboard when it seemed no one would come.

One infantryman summed it up later: "We thought we had been written off. Then the slicks came."

☐ **Primary Sources:** Mission logs and aviation records, Company A, 229th Assault Helicopter Battalion, 1st Cavalry Division, October–November 1965; 1st Cavalry Division After Action Reports, Pleiku Campaign.

☐ **Official Histories:** U.S. Army Center of Military History, *The Pleiku Campaign* (CMH Monograph, 1967); John M. Carland, *Stemming the Tide: May 1965 to October 1966* (CMH, 2000).

☐ **Published Works:** Harold G. Moore and Joseph L. Galloway, *We Were Soldiers Once... and Young* (Random House, 1992); Shelby L. Stanton, *Vietnam Order of Battle* (1981).

☐ **Veteran Accounts:** Vietnam Helicopter Pilots Association oral histories describing extractions near Plei Me (Capt. Thomas Rickman, WO1 Robert "Bob" Taft, WO1 Jack Swain) and the evacuation flights from Landing Zone Albany (Capt. Bill Beck, WO1 Bill Garay). Infantry recollections from survivors of Ia Drang recorded in VHPA interviews and 1/7 Cav veterans' associations.

☐ **Supplemental Sources:** Aircraft damage reports from A/229th AHB, November 1965; pilot award citations connected to the Pleiku and Ia Drang operations.

Dak To: The Hill of Fire

In late 1967, the Central Highlands became the scene of some of the fiercest combat of the Vietnam War. The town of Dak To was situated near the tri-border area where South Vietnam pressed againstmet Laos and Cambodia. The ridges surrounding it overlooked infiltration routes from the Ho Chi Minh Trail. For the North Vietnamese Army, seizing the high ground would open corridors deep into the south. For the Americans, losing it would cut off the Highlands and threaten the Central Plateau.

To defend the area, U.S. commanders committed heavy forces: the 173rd Airborne Brigade, elements of the 4th Infantry Division, Special Forces units, and battalions of the ARVN Airborne Division. Intelligence estimated that three NVA regiments had slipped into the region. It was clear that Dak To would be a fight on a scale not seen since Ia Drang in 1965.

The Highlands and Their Burden

The terrain punished men and machines alike. Ridges rose sharp and narrow, clothed in triple-canopy jungle that blotted out the sun. Valleys were thick with elephant grass taller than a man. Trails twisted through red clay that turned to glue in the rain. For infantry, every movement uphill meant hauling oneself up roots and vines, every bootstep a battle against the mud. For helicopters, the land was treacherous. There were a few open spaces, and those were obvious targets. Approaches were narrow, predictable, and exposed.

Pilots knew that every landing zone was an ambush waiting to happen. One Dustoff crewman later said that flying into Dak To felt like "running down a trench where everyone knows you're coming."

The Battle Begins

In early November, U.S. and ARVN battalions began sweeping the ridges around Dak To. Almost immediately, they met heavy fire. The NVA had prepared bunkers reinforced with logs, dug into slopes, and linked for crossfire. Every advance triggered ambushes. Artillery hammered the hills, and air strikes turned jungle into scorched stumps, but the enemy held.

Casualties mounted. In one week alone, hundreds of Americans were killed or wounded. Helicopters became the only lifeline. Huey slicks delivered reinforcements onto the ridge lines. Chinooks slung artillery pieces into valleys. Gunships circled constantly, rockets and miniguns raking bunker lines. And always there were the Dustoff birds, painted with red crosses, flying unarmed into kill zones to haul men out.

Hill 875

The climax came on Hill 875, a steep, jungle-covered rise southwest of Dak To. On 19 November 1967, the 2nd Battalion, 503rd Infantry Regiment, moved against the hill. A fortress met them. NVA troops occupied bunkers so well built that artillery and bombs barely dented them. Interlocking fields of fire cut into every approach.

The paratroopers fought uphill under steady fire. Mortars fell in deliberate patterns. Machine guns rattled from concealed pits. The jungle shook with explosions. By nightfall, the battalion had suffered devastating losses.

For three more days, the battle raged. The cost was horrific: more than 280 Americans killed, and hundreds wounded. Hill 875 became the bloodiest fight in the history of the 173rd Airborne Brigade.

For the helicopter crews, it was a gauntlet of courage.

The First Runs

On the morning of the 19th, Dustoff Hueys circled over the jungle, waiting for colored smoke to mark safe approaches. When green smoke popped through the canopy, the first pilot dove in.

The landing zone was little more than a scar in the trees, barely wide enough for a Huey. Fire opened instantly. Tracers cut across the clearing. Mortar rounds shook the ground.

The pilot flared nose-high, one skid scraping against the slope, while the other hanging in space. The crew chief leaned out, waving paratroopers forward. Stretchers came uphill, carried by men bent low under fire. Casualties were shoved across the skid and dragged inside. The medic pressed dressings against wounds as bullets punched holes through the fuselage. The door gunner fired bursts into the treeline, brass bouncing across the deck.

With the cabin loaded, the pilot eased off the collective. The Huey shuddered, reluctant to lift. Then the rotors caught and the helicopter staggered upward, tail boom rocking as mortar blasts chased it away.

Overloaded and Under Fire

That afternoon, another Dustoff came in already carrying wounded. More men waited. The crew could have abandoned the mission, but the pilot chose to land again.

Paratroopers carried more casualties forward until the cabin overflowed. Men lay shoulder to shoulder on the floor. When there was no more room, others clung to the skids, arms locked around struts. The crew chief tied one man to the door frame with a rifle sling to keep him from falling.

The medic crawled across the floor, trying to reach each casualty. At one point, he pressed a bandage with his elbow on one man while using both hands on another. The pilot pulled the power. For a long second, the skids dragged sparks from the slope. Then the helicopter lurched free, rising heavy, overloaded, but still airborne.

Chinooks in the Valley

By the third day, casualty stations overflowed. Commanders called for CH-47 Chinooks to move dozens of wounded at once. The big helicopters thundered into the valley, their rotors shaking the air.

Door gunners sprayed bursts into the jungle while stretchers were carried up the ramp. Crew chiefs lashed men to the deck with commo wire when straps ran short. One crewman recalled looking across the cabin and seeing a carpet of bodies, boots tangled, faces pale, the medic crawling from one to the next.

Each flight was a target. Tracers climbed after the Chinooks as they clawed skyward. Several returned riddled with holes, leaking hydraulic fluid, and with shredded panels. But they brought their loads back.

Voices from the Ground

For the men of the 173rd, helicopters meant life.

One paratrooper, bleeding from a leg wound, remembered lying in a crater with mortars still falling. "I thought I was gone," he said later, "but then I heard blades overhead. I knew I had a chance."

Another recalled being shoved onto a skid, his hands locking around the strut as the Huey lifted. "I didn't care if I fell," he admitted. "I just knew I was leaving that hill."

Crew members carried their own memories. A gunner said he fired until the barrel of his M-60 glowed red, empty brass piling around the wounded at his feet. A medic remembered tying a casualty upright against the cabin frame with a rifle sling because the floor was already packed. A crew chief spoke of scrubbing blood and mud out of the Huey with kerosene rags, only to launch again before the smell cleared.

Aftermath

When Hill 875 finally fell, the price was staggering. More than 280 Americans were dead, hundreds more wounded. The hill itself was a ruin — trees shredded, bunkers smashed, ground churned by artillery. The NVA had slipped away, leaving their dead in trenches.

For the 173rd Airborne Brigade, Dak To became a symbol of sacrifice and valor. For the crews that flew mission after mission, it was a test of endurance. Dustoff pilots flew into fire again and again, losing aircraft and men but saving countless lives. Slick pilots brought reinforcements onto ridges that seemed impossible to reach. Chinook crews hauled entire loads of wounded, pushing their machines to the limit.

One paratrooper summed it up years later: "We lived because the helicopters didn't quit."

Reflection

Dak To was a grim preview of what was to come at Hamburger Hill and other ridge fights. It showed the terrible cost of attacking fortified slopes in the jungle. It also showed the absolute necessity of helicopters. Without them, the battalions at Dak To might have been wiped out.

The Dustoff creed — *"When I have your* wounded" was never more accurate than on those ridges. The helicopters that clawed into Hill 875 turned what could have been a massacre into a bloody, costly victory.

Notes – Dak To

- Primary Sources: After Action Reports, 173rd Airborne Brigade, November 1967; 4th Infantry Division operational summaries, Dak To Campaign.

- Official Histories: U.S. Army Center of Military History, *Vietnam Studies: Tactical and Material Innovations*; Shelby L. Stanton, *Vietnam Order of Battle*.

- Published Works: John M. Carland, *Stemming the Tide* (CMH, 2000); Edward F. Murphy, *Dak To:*

America's Sky Soldiers in South Vietnam's Central Highlands.

- Veteran Accounts: Vietnam Helicopter Pilots Association oral histories of Dustoff and slick crews at Dak To; paratrooper recollections in the 173rd Airborne Association newsletters.

CHAPTER TWELVE

Khe Sanh: Helicopters Under Siege

I n January 1968, the Marine garrison at Khe Sanh Combat Base found itself under the eyes and the guns of the North Vietnamese Army. The outpost sat on a plateau in the northwest corner of Quang Tri Province, close enough to the Laotian border that the hills to the west ran all the way into Laos. Marines who arrived by air said the approach could look beautiful, even peaceful, but once they were on the ground, it felt like they were surrounded and boxed in by terrain that handed the enemy the high west seemed to lean into another country. To the east, the land opened into rolling red-clay ridges, dotted with scrub and scarred by the remains of old French fortifications. The western side was another world entirely, a dark wall of jungle and steep mountains that ground.

The 26th Marine Regiment manned the defenses, stretched around the airstrip and the outlying strong points. They lived in bunkers dug deep into the clay and in trenches that zig-zagged across the perimeter. The North Vietnamese Army had massed two divisions outside the wire and knew precisely where those trenches lay. Shells fell every day, sometimes every hour. Mortars and rockets walked across the base in deliberate patterns, collapsing bunkers, flipping trucks, and scattering men who had no place to hide. When the barrages slackened, the smell of burned powder and churned clay hung in the air like fog.

Inside the wire, the Marines fought to keep the base alive. The airstrip was their only real artery, and it was cratered repeatedly, leaving sections of it looking like a patchwork quilt of filled-in holes. C-130s and C-123s attempted to land with supplies, but the runway was often too damaged or exposed to support safe landings. That left the helicopters.

The CH-46 Sea Knight and the larger CH-53 Sea Stallion became the workhorses of Khe Sanh.

They flew in ammunition, rations, replacement barrels for the artillery, and mail that smelled of home. More importantly, they evacuated the wounded. Marines remembered the aid stations overflowing within the first week of the siege. Trenches became triage lines, with ponchos spread on the mud to hold men waiting for a lift. Corpsmen worked by feel in the dark, bandaging and giving morphine under the flicker of lanterns while the artillery boomed outside.

Every evacuation was a gamble. To reach the helicopters, wounded men had to be carried across open ground while shells still fell. When the rotors thumped overhead, stretcher teams ran through smoke and dust, heads down, boots slipping in the clay. Door gunners fired into the tree line, trying to keep the enemy's head down for a few more seconds. The cabins filled fast, litters stacked two high, Marines slumped against bulkheads with bandages already soaked through.

The siege of Khe Sanh lasted 77 days. In that time, helicopters flew thousands of sorties. Dozens were shot down or destroyed on the ground. Crews knew the odds every time they launched, but without them, the base could not have held. The Marines in the trenches learned to measure hope by the sound of rotors beating through the smoke.

The First Run – February 1968

The men in the trenches around the runway heard the heavy rotors long before they saw the helicopter. The sound grew into a pounding that seemed to rattle the clay walls of their bunkers. Through the haze and drifting smoke, a CH-46 Sea Knight appeared, its twin rotors biting at the air, nose tilted slightly down as it lined up for the strip. At the controls sat 1stLt Ray Stout, eyes locked on the broken pavement, with 1stLt Paul Jensen beside him working the radios and watching the gauges.

The ridges to the north and west lit up at once. North Vietnamese crews had waited for the moment a bird came into view.

Tracers slashed across the gray sky, and mortars began to fall in sequence. One round hit close enough that shrapnel clattered against the nose, leaving scars in the metal. "Keep her steady," Jensen called over the intercom, his voice tight but measured. Stout nodded, hands firm on the controls, easing the Sea Knight down. The wheels smacked the cratered strip hard enough to jolt helmets.

In the back, Sgt. Edward Fox, the crew chief, was already unbuckled, leaning into the doorway. "Let's go, let's go!" he bellowed, waving stretcher teams forward. Marines sprinted through the haze, boots splashing in water-filled craters, heads down as shells burst along the edges of the strip. Two men carried a Marine whose trouser leg ended above the knee, blood soaking through the poncho they used as a litter. Fox hauled him in with both hands, shouting for the medic to apply a tourniquet quickly.

On the other side, Cpl. Alton Slaughter leaned into his M-60, firing long bursts toward the tree line. The barrel heated and steamed in the damp air, tracers arcing out into the brush. "More coming!" he shouted as another squad pushed forward with two litters, the cabin filled with wounded in minutes. Litters were wedged along the deck, Marines slumping against bulkheads with bandages darkening by the second. One casualty gasped for air, his chest wound bubbling with each breath. A corpsman dropped beside him, hands already slick with blood, calling out for morphine.

"Full load!" Fox yelled, pounding the bulkhead with his fist. "We're packed tight." Stout glanced once at Jensen, who gave a quick nod, then eased back on the collective. The Sea Knight lifted heavy, nose dipping as if reluctant to leave the ground. For a moment, the wheels dragged across the torn runway, as if an invisible hand was holding them to the ground. Then the rotors caught clean air, and the bird surged upward.

Mortars chased them as they climbed. One blast off the port side rocked the helicopter hard, throwing Fox against the bulkhead and knocking helmets together in the cabin. Slaughter stayed on the gun, firing until the ammo belt was consumed, then slapping another into place. "Tail's smoking," Jensen reported, reading gauges through the haze. Stout kept her climbing, every movement smooth, coaxing the wounded machine to stay aloft. Behind them, the runway was still under fire, but inside the helicopter, thirty Marines were alive who otherwise would have bled out in the mud.

The Second Run – The Sea Stallion

The Marines called it the "big bird," and when the CH-53 Sea Stallion dropped into Khe Sanh, the ground seemed to shake with its arrival. Its twin engines threw up a storm of red clay and dust, rotors chopping the air so hard men in the trenches felt it in their chests. At the controls was Maj. William J. White of HMH-463, with his copilot, Capt. George McKinney beside him. They had flown heavy transports all across I Corps, but nothing compared to the gauntlet of Khe Sanh.

The approach was slow and steady. The Stallion was massive and noisy, and every gunner in the hills knew it was coming. Tracers streaked up from the north, and mortar rounds bracketed the strip. The helicopter rattled as bullets punched through panels, a metallic hammering that echoed through the cabin. White kept his hands steady, easing the nose down. "Hold her, hold her," McKinney called, scanning gauges. The wheels slammed onto the torn runway, jolting the whole airframe.

The rear ramp dropped, and Sgt. Richard Vance, the crew chief, waved stretcher teams forward. Marines came at a run, carrying litters containing bleeding Marines, dust whipping into their faces as they bent low against the fire. Boots skidded and sank in the churned clay, and more than one man nearly fell before reaching the ramp. Inside the cavernous helicopter, the crew wrestled with the stretchers, sliding them across the deck and stacking them where they could. The space filled quickly, bodies pressed in side by side, boots knocking together, bandages turning dark and wet. One Marine groaned with every breath, a ragged sound that carried even over the engines. Another lay still, eyes fixed on the ceiling, lips working silently as if in prayer.

When the regular straps ran out, Vance grabbed a spool of commo wire and lashed litters to the deck. "Make it hold, just make it hold," he muttered as he tied knot after knot. At the side windows, door gunners fired into the tree line, brass spilling across the floor, the smell of hot barrels mixed with fuel fumes and blood.

More Marines came running, dragging their wounded on ponchos. The ramp became a bottleneck. Vance shouted for them to keep moving, pulling men inside by their web gear. A sergeant pushed three men forward and then collapsed himself, bleeding from a head wound. They hauled him in, too. Within minutes, the Stallion was jammed with wounded men.

"Load is heavy," McKinney warned, watching torque climb. White nodded, face set. "We go anyway." He raised the collective. The big bird groaned as if it would stay nailed to the runway. For a long second, the wheels dragged, as if the great helo was mired into the runway. Then it was as if the rotors bit clean air and the Stallion rose, ramp still down, dust and debris swirling in its wake.

Enemy fire chased them. A mortar round landed on the strip just as they cleared it, the shockwave slamming the tail. The helicopter lurched, Marines inside clutching stretchers to keep them from sliding. Vance braced himself on the ramp, one hand on a frame, eyes on the men laid out before him. He thought more than once the machine would not make it.

But White coaxed the big helicopter higher, every movement deliberate, fighting for altitude. The bird climbed unsteadily, tail yawing, but it stayed in the sky. When they finally cleared the ridges, the crew chief slumped against the bulkhead, exhausted. The cabin behind him looked like a battlefield, bodies lashed with wire, Marines moaning, corpsmen working desperately as the helicopter droned on toward the aid station at Dong Ha.

Dozens of men who would have died in the mud at Khe Sanh lived to fight another day because the Stallion lifted when it had no right to.

The Third Run – A Sea Knight Shot to Pieces

The next Sea Knight came in low out of the overcast, a Marine bird from the Purple Foxes, HMM-364, its paint already pocked by earlier runs. From the ridge north of the strip, the guns found it quickly. Tracers rose in a steady ladder, and mortars began to feel their way across the runway, one burst at a time. In the cockpit the aircraft commander kept his hands easy on the controls, talking softly into the intercom, counting off distance markers that were half buried in craters. His copilot handled the radios and watched the needles, torque flirting with the red as they bled off airspeed.

They took the first hit crossing the wire. A round slammed into the nose and showered the cockpit with fragments. The pilot flinched, blinked the grit out of his eyes, and nosed the helicopter down toward the pitted concrete. Another burst rattled the forward door, a hard metallic slap that everyone felt through the seats. "On the ground," the copilot called, though the skids were not truly level, one biting deeper than the other where a near miss had torn the runway. The whole machine sat at a tired lean, working to keep itself upright.

In the cabin, the crew chief kicked the ramp release and felt the big door start down, its hinges grinding under a coat of red-clay dust. He was already waving stretcher teams forward. "Bring them, bring them," he yelled, voice raw from smoke and turbine howl. Marines ran bent low, four hands on each litter, boots slipping in the clay. The first two casualties were on the ramp before the door settled all the way, corpsmen grabbing shoulder straps and hauling them inside. His web gear dragged a third, his trousers torn, and one boot was missing. The door gunner at the starboard window leaned into his M-60 and poured a long burst into the tree line, brass clattering across the deck plates and rolling under the litters.

Mortars walked closer. One round hit the edge of the strip and the concussion slapped through the helicopter, knocking helmets against bulkheads and dropping a haze of dust from the ceiling panels. The crew chief grabbed the ramp stanchion to steady himself and then bent over a Marine whose chest rose in short, sharp pulls. Blood bubbled around a dressing that would not hold. "Tape," he shouted, not looking up. The corpsman tore a strip with his teeth and pressed down hard; fingers slick.

More stretchers came. The space filled faster than it should have. Litters were stacked along the centerline. Men slumped on the deck where there was no room left, backs to the bulkheads, eyes closed, faces gray with shock. "That is it," the crew chief called, pounding the ramp with his fist. "We are full." He turned to wave off another team and saw a Marine stumbling toward him with a man over his shoulder, both lurching with each step. The crew chief grabbed the casualty's belt and pulled; the corpsman caught an arm; together they slid the man across the ramp into the crush of bodies.

The pilot felt the weight in the controls, heavy and slow. "Power coming," he said into the mic. The copilot gave him gauges and a quick, quiet "You are good." He lifted the collective a little at a time. The helicopter shivered but did not move. He let it settle, tried again. The wheels scraping across broken matting, sparks flashing in the dust. The gunner fired a short string to keep heads down. "Go, go," the crew chief barked, one hand on the ramp frame, eyes never leaving the men sprawled at his feet.

They started to rise, inches at first, the oversized rotors digging for something cleaner than the chopped air pressed against the strip. Another mortar burst went off to port, and the Sea Knight rolled hard. For a breath the pilot thought they would tip. He pushed in a little right pedal, smoothed the roll with light hands, and felt the helicopter steady. "Climb," the copilot said, not quite a prayer. The ship obeyed, only just.

Halfway down the strip, a burst of machine-gun fire stitched the tail. The entire airframe shuddered, and the pilot smelled the sharp, chemical tang of hydraulic fluid, cutting through the smoke and blood and fuel. Warning lights flickered along the panel, small red eyes staring at the flight crew. "We are leaking," the copilot said. "Copy," the pilot answered, and kept his nose down to build airspeed. In the cabin, the crew chief shouted at a corpsman to hold a litter upright as the deck tilted. The gunner slapped in a fresh belt and squeezed a short burst, then another, not wasting rounds.

They cleared the runway and drifted across the perimeter trenches. Marines below looked up and saw the Sea Knight pass, ramp still half open, dust and paper whipping in its wake. Inside, a man groaned with each jolt and then fell quiet, and the corpsman leaned close to check his breathing, hand flat against a bandage that trembled with every heartbeat. The crew chief slid along the ramp on one knee and cinched another length of commo wire around a litter frame, knotting by feel.

The plateau fell away. The helicopter yawed left as the pilot trimmed for the climb, and the tail stopped shaking. The lights on the panel steadied as well. It was not a clean aircraft anymore, but it was flying. The pilot gave the engine a little more, careful not to spike the needles, and headed for Dong Ha, a slow arc across the hills. No one in the cabin said much. The gunner watched the tree line until the ridges slipped behind. The crew chief bent over the men on the floor and spoke to them one at a time, telling them they were aboard and to hold on. In that moment the Sea Knight felt like the only dry place left in the world.

When they landed at the aid station, the crew would count the holes, and the mechanics would shake their heads. For now, the only count that mattered was the number of Marines still breathing as the helicopter droned on through the gray afternoon.

Voices of Marines

For the men who fought from the trenches and craters of Khe Sanh, the sound of helicopters meant everything. Mortars and rockets hammered the base day and night, shaking the ground until it felt like the whole plateau would buckle. But when the thump of rotor blades broke through the barrage, it carried a promise. Helicopters meant food, water, ammunition, and most of all, a ride out for the wounded.

One Marine from Bravo Company, 1/26, later recalled lying in a trench line with his arm wrapped in a bloody bandage that barely held. "I could hear the artillery, the rockets coming in, and I remember thinking, this is it, we're not getting out of here. Then I heard that chop, chop, chop overhead. You knew what it was right away. That sound meant you had a chance." He said the memory stayed with him long after the war, the sudden rush of hope in the middle of despair.

Another survivor remembered sprinting across the open strip with three other men carrying a stretcher. He remembered the way the ground bucked under the blasts, each concussion slapping his boots as he ran. Dust and smoke blew straight into his face, and he could hardly see the stretcher in front of him, just the outline of the man they carried. All he thought about was reaching the helicopter that waited in the haze, the only thing solid in the chaos. Years later, when he was asked about it, he shook his head and said, "It felt like running through hell, and that bird was the only chance our buddy had." He said he had no memory of the gunfire, only of the sight of the crew chief waving them in, hands cutting through the dust cloud.

The aircrews who flew the missions carried their own memories of those days. A door gunner with HMM-262, then only nineteen years old, recalled what it felt like to lean into his weapon while shells rained down on the strip. "You're firing blind most of the time, just laying tracers into the tree line. But you know the grunts are watching you, and they depend on you to keep heads down. So, you don't stop." He remembered going through three barrels in one flight, the metal glowing red as he swapped them out with shaking hands.

Crew chiefs had the worst of it, half in and half out of the helicopters. They hauled litters, dragged wounded by their web gear, and shouted orders nobody could hear over the turbines and the explosions. One sergeant later said, "The smell is what I never forgot, blood, oil, smoke, all mixed. You couldn't get it off your gear. You couldn't get it out of your head." He had to wipe the deck plates clean with rags and kerosene after each flight, only to see them covered again within an hour.

Pilots spoke with the same steady tone they used in combat, but the strain came through in their words. A CH-53 aircraft commander said bluntly, "It was like flying into a meat grinder. Every time you made that approach, you knew you might not lift off again. But the Marines down there were counting on you. That was enough." He described the heavy silence in the cockpit after landing back at Dong Ha; nobody spoke, each man just sat for a moment, breathing, until someone broke the quiet with a nervous laugh or a curse.

To the infantry, the helicopters were angels of mercy. To the crews, they were missions that had to be flown, no matter the odds. Between them, a bond was formed that outlasted the war itself. Marines who walked out of Khe Sanh alive never forgot the sight of a Sea Knight or a Sea Stallion breaking through the smoke. Crews who flew in swore they could still hear the pounding of the rotors decades later.

Aftermath

The siege of Khe Sanh lasted seventy-seven days. It felt longer to the men who endured it, nearly three months of incoming fire, sleepless nights in bunkers, and the constant rattle of helicopters overhead. When the last North Vietnamese shells landed in April 1968, the Marines were still holding the wire on paper that counted as victory. Official tallies listed thousands of enemy dead, hundreds of Marines killed or wounded, and mountains of ammunition poured into the surrounding hills. Those numbers filled reports, but they only hinted at what the place had been like. For the Marines in the trenches, the days blurred together into a cycle of shell bursts, carrying the wounded to the strip, and watching helicopters lift out overloaded with men who might or might not survive the flight.

The helicopters had made the difference. Without them, the garrison would have run out of everything that mattered: food, water, artillery rounds, bandages. They were the thread that kept the base tied to the outside world. Crews flew into an airstrip that was never quiet, often cratered, and constantly watched by enemy gunners. They lifted loads far beyond regulation, decks stacked with litters, Marines jammed shoulder to shoulder, some hanging on to the ramp as the big machines clawed into the sky. More than forty helicopters were destroyed during the siege, and many more came back with holes stitched across their skins. Mechanics patched what they could and sent them out again, because the Marines on the line had no other lifeline.

For a time, Khe Sanh carried enormous weight back home. President Johnson had maps of the place spread across tables in the White House. Journalists called it another Dien Bien Phu in the making. When the siege lifted and the base still stood, it was hailed as proof of American endurance, an outpost held against overwhelming odds. To many, it showed that U.S. firepower and mobility, anchored by helicopters, could blunt the enemy's boldest moves.

Yet the victory rang hollow. Within months, the Marines were ordered out. By July 1968, the airstrip was empty, bunkers collapsing into the clay, weeds spreading across the landing zones where Sea Knights and Sea Stallions had once fought their way skyward under fire. The decision looked sensible in the planning rooms: the base no longer served the strategy, and the enemy had shifted elsewhere. But for the men who had dug in, who had carried friends to helicopters that barely cleared the hills, it was harder to accept. One Marine put it simply years later: "We held that ground, and we paid for it. Then we walked away like it never mattered."

Khe Sanh proved the courage of the helicopter crews and the toughness of the Marines who endured the siege. It also revealed the contradictions of the wider war — battles turned into symbols, then abandoned when strategy changed. The helicopters had saved a garrison, but they could not save the meaning of the place.

☐ **Primary Sources:** Command chronologies of the 26th Marine Regiment, January–April 1968; 1st Marine Aircraft Wing aviation records; official U.S. Army and Marine Corps after-action reports from the siege of Khe Sanh.

☐ **Official Histories:** U.S. Marine Corps, *U.S. Marines in Vietnam: The Defining Year, 1968* (Historical Branch, Headquarters USMC, 1997); U.S. Army Center of Military History monographs on the Tet Offensive, 1968.

☐ **Published Works:** Ray Stubbe and John Prados, *Valley of Decision: The Siege of Khe Sanh* (Houghton Mifflin, 1991); Shelby L. Stanton, *Vietnam Order of Battle* (1981).

☐ **Veteran Accounts:** Oral histories from CH-46 and CH-53 pilots and crews preserved in the Vietnam Helicopter Pilots Association archive; U.S. Naval Institute oral history program interviews with HMM-262 and HMH-463 veterans; testimonies of Marines evacuated under fire recorded in Khe Sanh Veterans Association publications.

☐ **Supplemental Sources:** Aircraft loss and damage reports, 1st Marine Aircraft Wing, 1968; contemporary reporting from *Stars and Stripes* and Associated Press dispatches during the siege.

Chapter Thirteen

Hamburger Hill: Lifting Men from the Slopes

In May 1969, the A Shau Valley once again became the center of a brutal fight. The valley, tucked against the Laotian border, was an extended, narrow cut through the mountains, green and thick with jungle, heavy with mist in the mornings. It had been a North Vietnamese stronghold for years, a natural corridor for men and supplies flowing down the Ho Chi Minh Trail. American units had gone in before and paid a heavy price. In the second week of May, soldiers of the **101st Airborne Division** went in again, their orders to seize Hill 937, one of the ridges that dominated the valley.

Hill 937 rose steep and unforgiving, covered in elephant grass and scrub trees near the top, tangled jungle lower down. It did not look like much on a map, just another numbered rise among dozens, but to the men who climbed it, it felt endless. The rains came and turned the slopes into mudslides. Soldiers clawed at roots and vines for grip. Packs slid down behind them, weapons clogged with grit. Every step upward was met with fire. The **29th North Vietnamese Regiment** was dug in across the crest, their bunkers reinforced with logs and earth, interlocked to cover each approach.

The battle opened on May 10. Infantry companies pushed up the slope in staggered assaults. The pattern became grimly familiar. Soldiers moved out of the tree lines, rifles held low, boots sinking in the muck. Mortars and machine-gun fire tore into them from hidden positions above. Men fell, others dragged them back, and still the units tried again. By the third day, the hill had already earned its name. Soldiers of the **3rd Battalion, 187th Infantry**, the Rakkasans, said it was chewing men up like meat in a grinder. "Hamburger Hill" became the word that stuck.

Helicopters orbited overhead from the beginning. They carried resupply, ammunition, water, rations, but most of all, they had the wounded back down. The hill offered almost nothing like a landing zone. Pilots flared their Hueys onto slopes so steep that one skid dug into the mud while the other hung in the air. Door gunners leaned out, firing into the jungle, while crew chiefs pulled casualties up by their web gear. Medics worked in the open, dragging men through muck to get them aboard.

Sometimes soldiers lashed the wounded to bamboo poles and shoved them onto the deck as the helicopter rocked under fire.

The weather was a fight all its own. Most afternoons, the rain came in sheets, hammering the ridges until the whole valley blurred gray. Low clouds hung stubbornly over the A Shau, so the pilots hugged the hillsides, following ridgelines and skimming treetops to stay oriented. Rotor wash whipped the muck into the air, turning it into grit that peppered eyes and bare skin. Inside the cabins, it was close and foul, the stink of wet earth mixed with sweat, gunpowder, and hydraulic fluid that never seemed to wash out. Crews said later that the conditions were some of the worst they had ever faced, yet they flew sorties all day, every day, because without them the wounded would never leave the slopes alive.

By May 20, after ten days of climbing, falling back, and climbing again, Hill 937 was finally taken. The price was staggering: **72 Americans dead, 372 wounded**. The helicopters had lifted nearly all of those casualties off the hill, some in daylight, some at night under flares, constantly under fire. Almost every wounded man who lived to tell of Hamburger Hill owed it to the helicopters. They came in daylight when they could, at night under parachute flares when they had to, and they never came without drawing fire. Among the grunts, the talk turned half-serious, half-bitter, that without the birds they would have been left in the muck until the rain covered them. The laughter that followed carried a hard edge, because deep down each man knew it was true. Those helicopters were the line between getting home and being another body lost on a nameless ridge.

The First Runs – Dustoff on the Slope

The call came on May 15, 1969, midway through the battle for Hill 937. Companies of the 3/187th Infantry had clawed up the slope all morning, only to be shredded by machine-gun and mortar fire. Casualties were piling in shallow depressions and behind tree stumps, bandaged as best they could be by medics working under fire. The battalion aid station was overflowing. If the wounded were not pulled off the slope, many would not live until nightfall.

The mission fell to a Huey from the **326th Medical Battalion**, call sign **Dustoff 65**, flown that day by **CW2 Robert H. Garwood** with **1LT John Stiles** in the left seat. In the back were crew chief **SP4 James D. Nichols** and medic **SP5 Larry Wood**. They had already made three sorties in the valley that week. None had been as desperate as this one.

Garwood nosed the Huey along the valley floor, weaving through clouds that clung to the ridges. Rain streaked the windshield in sheets. Stiles kept one eye on the gauges and another on the ridgeline ahead, calling out headings. "Ridge in ten seconds, low left," he warned. Garwood banked slightly, hugging the slope until the muddy clearing opened ahead. They could see muzzle flashes sparking from bunkers at the crest. "We're taking fire," Nichols shouted over the intercom as tracers hissed past the tail.

Garwood flared, fighting to hold the Huey steady against the incline. One skid bit into the mud while the other hovered a foot above the slope, the helicopter rocking in the wind. Nichols leaned out the door, one boot braced on the skid, and waved frantically at the infantry below. "Bring them up, now!" he yelled, his voice torn away by rotor wash and gunfire.

Four paratroopers scrambled uphill, bent low, carrying a man slung in a poncho. They slipped and skidded, mud up to their knees, before reaching the helicopter. Nichols and Wood grabbed the casualty by his web gear and hauled him across the skid into the cabin. The man groaned once and went limp. Wood dropped beside him, tearing open bandages and pressing hard against the wound. Another team of soldiers appeared, dragging two more wounded between them. Stiles twisted in his seat to glance back, his face pale in the glow of the instruments. "We're heavy," he muttered.

Outside, enemy fire hammered into the slope. Rounds smacked the fuselage with sharp metallic thuds. "We're getting stitched up," Nichols called, jerking back as a burst of tracers whipped across the doorway. He turned and dragged another man in, mud and blood smearing the deck plates. The cabin was already crowded, litters stacked two high, soldiers slumped against the bulkheads with bandages darkening by the second.

"Full load!" Nichols yelled, pounding the bulkhead. Garwood nodded once, hands locked on the controls. He eased the collective, coaxing the helicopter upward. For a moment, the skid buried in the slope held them, the Huey shuddering as if it would tip sideways. Then the blades caught clean air, and the ship lurched into the sky. Stiles called out torque readings, his voice tight. "Watch it, watch it."

Mortars bracketed the clearing as they lifted. One burst below sent a jolt through the helicopter, rattling helmets and throwing Wood against the bulkhead. He steadied himself and bent back to his patient, shouting for morphine. Nichols braced in the doorway, rifle slung, eyes on the jungle as tracers reached for them. "Come on, come on," he muttered, as if speaking to the Huey itself.

The helicopter staggered across the treetops, tail streaming smoke. Garwood coaxed it along, every movement careful, until the ridgeline slipped behind. Only then did Stiles exhale and lean back in his harness. Inside the cabin, Wood kept working, his hands slick and trembling. Nichols looked at the pile of wounded at his feet and shook his head. They had lifted eleven men off that slope in one run, far more than the Huey was ever meant to carry.

The Second Run – A Huey Under Fire

Later that same day, another call went out from the slopes of Hill 937. A company of Rakkasans had been caught in crossfire as they tried to push through thick brush below the crest. Several men were down, pinned near a tree line cut to ribbons by machine-gun fire. The battalion begged for a Dustoff.

The mission fell to **1LT Charles "Chuck" Brown**, flying a UH-1H with **CW2 Tom Harris** as his copilot. In the back were crew chief **SP4 Dennis York** and medic **SP5 Michael "Doc"**

Jensen. They had been circling the valley for nearly an hour, watching bursts of tracers run across the slopes, waiting for a break. Finally, Brown pushed the nose down, determined to get in before the weather closed again.

The approach was ugly from the start. Rain slicked the canopy, and the clouds hung low, pressing the valley into shadow. Harris called headings, his voice sharp over the intercom. "Left five... ridge dead ahead." Brown kept her steady, eyes flicking between the gauges and the brown scar of mud where he meant to plant the skids.

The enemy opened up the moment they saw the Huey. A burst of fire stitched across the nose, shattering Plexiglas and spraying the cockpit with fragments. Harris ducked, blood on his cheek. "I'm good, keep flying," he barked. Another burst struck the fuselage near the fuel cell, causing the whole aircraft to jolt with the impact. In the back, York shouted, "We're hit, we're hit!" but kept leaning out to wave stretcher teams forward.

Brown flared hard, trying to hold the ship steady against the slope. One skid was buried deep in the mud, while the other hung in space. Soldiers came scrambling up, dragging two wounded by their harnesses. York and Jensen grabbed arms and webbing, hauling the men inside. One casualty screamed with each jolt, another barely moved. Jensen dropped beside them, checking their airways and ripping open bandages. "He needs morphine now," he yelled, though no one could hear him above the turbines.

More soldiers pushed forward, shoving two more wounded into the cabin. Blood smeared the deck plates, mixing with rain and mud. York braced himself, one boot on the skid, and pulled another Marine aboard by the straps of his flak jacket. "We're packed!" he hollered.

The Huey shuddered as another burst tore through the tail boom. Warning lights flickered red across the panel. Harris glanced at the gauges. "We're losing hydraulics." Brown's voice stayed steady: "We're not leaving them." He gave a sharp nod to York, who slammed the door and signaled clear.

Brown eased the collective. The helicopter resisted, heavy with men. For a long second, the skids scraped through mud, the nose yawing left. Harris fed in the pedal and muttered a curse. Then the rotors bit and the Huey lifted, sluggish, every inch fought for.

As they cleared the slope, an RPG whooshed from the tree line. It missed by yards, passing under the belly and bursting harmlessly in the mud. The shockwave still rattled the bird, helmets knocking together in the cabin. Jensen sprawled across two casualties, shielding them with his body until the jolt passed. "Hold on, damn it," he shouted, hands pressed hard against a chest wound.

Brown nursed the helicopter along the ridgeline, nose dipping, tail smoking. Harris kept his eyes glued to the gauges, calling torque and temp numbers like a prayer. "She'll hold," Brown muttered, more to himself than to anyone else. At the back, York leaned out the door, watching tracers climb up after them, and thought the machine was too slow, too heavy. Yet somehow it climbed.

When they finally dropped into the aid station, the Huey looked like a wreck. Holes peppered the fuselage; one door was almost torn away, and hydraulic fluid was slick on the tail boom. The mechanics shook their heads and told Brown that the bird would never fly again. York and Jensen didn't care. They counted seven wounded alive on the floor, men who would have died in the mud if the Huey hadn't made the run.

The Third Run – Night Under Flares

By the fifth day, the hill barely looked like a hill at all. The top was stripped bare, trees reduced to jagged stumps, the slopes carved into slick slides of red mud by rain and artillery. Deadfall and broken gear littered the ground. The wounded lay wherever they had been pulled, some in hasty clusters behind logs, others still scattered where they had gone down. Darkness brought no pause, the shooting carried on, mortars thumped through the night, and men kept falling. North Vietnamese gunners fired through the darkness, and American artillery answered. The wounded still needed to be lifted off, and helicopters still came.

One of those flights belonged to **CW2 Daniel "Danny" Cole**, piloting a Huey with **1LT Mark Reynolds** as copilot. Their crew chief was **SP4 Thomas "Tommy" Laird**, and the medic was **SP5 Robert Scott**, both already worn down after three straight days of missions. They had never flown a night extraction under flares this close to an enemy bunker line.

Cole eased the Huey along the valley floor, hugging the dark line of trees. The only light inside was the faint green glow of the gauges; glances before his eyes jumped back outside. Above, the artillery had thrown flares into the sky. They hung on parachutes, drifting, swinging in the wind, each one flooding the ridge for a moment before the light slid away again. The ground seemed to change shape with every burst, mud shining wet, shadows stretching and twisting. From the cockpit, every stump, every shattered log looked like a man waiting with a rifle. To Cole, it felt less like flying into a battlefield than into a nightmare painted in light and shadow. Reynolds called altitude changes, his voice quick, keeping them from drifting into the hillside.

As they neared the clearing marked by green chemlights, fire opened up. Tracers cut red arcs through the flares' glow. "Taking fire, right side!" Laird shouted, leaning out with his rifle. Cole kept the nose steady. "We're going in," he said, voice flat. The Huey flared against the slope, skids uneven, one buried, the other hanging over emptiness.

Soldiers appeared in the half-light, struggling through the mud with stretchers. In the white glare of the flares, the stretcher teams looked unreal, helmets flashing as they slogged through the mud, faces drained and hollow. Laird leaned half out of the doorway, arms cutting the air to wave them forward. Inside the cabin, Scott was already on his knees, shoving gear aside to make room. "Get them in quick, we can't stay here," he shouted, his voice nearly lost under the turbines. The first two litters slid aboard, mud pouring off the ponchos, blood soaking through the fabric. Scott worked by muscle memory, hands moving from dressings to morphine syrettes, doing what he could while the helicopter rocked and bullets smacked the fuselage.

Another team staggered up with a man shot through the abdomen. They shoved him forward, and Laird grabbed him by the flak vest, pulling him inside. "He's in!" he shouted, but even as he spoke, another burst cracked across the fuselage, holes popping through the skin. Reynolds swore into the mic. "We've got hits left side." Cole kept his grip, holding the ship against the slope.

Inside the cabin, it was a scene of bedlam. Scott pressed a dressing against the wound with both hands, trying to stop the flow of blood. "Stay with me," he muttered, his words lost in the turbine howl. Laird crawled over two litters to lash another in place with commo wire. He felt the bird shiver as a mortar exploded somewhere behind them. Dust and grit rained down from the ceiling panels.

"Last one!" someone shouted outside. A squad carried a soldier with a spine injury, moving as carefully as they could in the muck. They pushed him in, faces tight with fear. Laird slammed the door. "We're packed!" he barked.

Cole eased the collective, the Huey groaning with the weight. For a moment, the bird refused to climb. Then the blades caught a clean pull of air, and she lifted, nose dipping forward as Cole pushed for speed. Fire followed them, tracers climbing, but the flares were burning out, and darkness closed in.

They cleared the ridge. Reynolds leaned back, his voice suddenly quiet. "We're out." In the cabin, Scott sat in blood and mud, still working, sweat running down his face. Laird looked around at the wounded sprawled on the floor and thought, not for the first time, that the only miracle was how many were still breathing.

Ground Voices

The men who clawed their way up and down Hill 937 remembered different things, but the mud came up in nearly every story. It coated everything, boots, rifles, bandages, and it never let go. A paratrooper from Alpha Company, interviewed years later, laughed once before saying it wasn't the gunfire that stuck with him; it was dragging a wounded buddy through that mess. "Every step he slid, I slid. We went down on our faces more than once. By the time we reached the bottom, we looked like we'd both been buried in the hill.". "You slipped, he slipped, and both of you ended up face-first. Then you tried again. By the time we got to the pickup point, we were caked so heavily you couldn't tell who was hit and who wasn't."

The helicopters gave them a reason to keep pushing. A staff sergeant remembered the first time he heard a Huey trying to settle on the slope. "You heard that whop, whop, whop coming closer, and we thought no way, no way she's landing here. Then she did. One skid sunk in, the other hanging out over nothing, and the crew chief waving like mad for us to hurry." He swore that sight, the bird holding steady while rounds cracked all around it, gave him the strength to take another step.

The crews had their own stories. A nineteen-year-old gunner with the 326th Medical Battalion said his job was simple: keep fire off the ramp. "I went through three belts in one run. Didn't see targets clearly, just laid tracers where the shooting seemed to come from. I knew the grunts were counting on me, so I didn't stop." He remembered the barrel glowing hot, smoke in his face, his hands shaking as he swapped it out for a cooler barrel.

Crew chiefs lived half in and half out of the helicopters. One sergeant talked about dragging casualties by their webbing, yanking them across the skid, and shoving them inside with his shoulder. What stayed with him was the smell. "Blood, oil, smoke, sweat, it's all mixed together. You never scrubbed it off. I wiped the floor with rags and kerosene, and an hour later it was back."

The medics remembered little but the wounded. SP5 Larry Wood said the floor was never dry. "Rainwater, mud, blood. It all ran together. You crawled from one man to the next while the bird shook, trying to hold bandages down. Sometimes I had a hand on two guys at once, hoping both would make it until we landed."

For the grunts, the helicopters were more than machines. They were a constant reminder that someone outside the ridge still remembered them and still thought their lives were worth the risk. For the crews, it was a job, brutal and constant, get in, pull them out, patch the ship, go again. Both sides agreed on one thing: the sound of those rotors was the only thing that cut through the rain, the fire, and the shouting, and told a man he still had a chance to live.

Aftermath and Reflection

By May 20, the powers-to-be had declared Hill 937 to be in American hands. The climb had taken ten long days. The cost was heavy, with 72 men killed and more than 370 wounded. The hilltop was littered with shattered trees, burned bunkers, and the smell of cordite and rain-soaked earth. The official reports listed the numbers clearly, with columns of losses and ammunition expended. The soldiers remembered something else: the weight of exhaustion in their bones, the constant rain on their helmets, and the helicopters that came repeatedly when they thought they would never leave that slope alive.

Dustoff and assault Hueys had lifted nearly every casualty off the mountain. Some missions were conducted during daylight under covering fire. Others came in at night under parachute flares, the jungle below lit up in strange, swaying light. Crews balanced their machines on angles no manual approved of, dragging men aboard in seconds, then fighting to coax the helicopters into the air while bullets hammered their skins. More than once, wounded were lashed to bamboo poles and shoved inside because there was no time or space for litters. Crews described it later as pure chaos — mud, shouting, the stench of blood and sweat, and then the sudden lift into the air that felt like a reprieve from a death sentence.

When the hill was finally taken, commanders hailed it as a victory. The enemy's 29th Regiment had been torn up, their bunkers destroyed. Yet the name "Hamburger Hill" followed the battle home. Reporters told stories of men ground down in hopeless assaults, of helicopters overloaded with casualties, of a hill abandoned only weeks later. The controversy spread quickly. Parents asked why their sons had died for a ridgeline the Army did not even keep. Soldiers who fought there asked the same question in quieter voices.

For the men who flew the lifts, all the arguments that came later seemed far away. What stayed with them were the faces on the deck, the weight of a man being dragged across the skid, the sound of somebody groaning behind the bulkhead. A pilot said years afterward, "That ship wasn't built for what we asked of her, but she did it, and because of that, a lot of boys made it home." The infantry who rode those birds remembered the same thing, not politics, not speeches, just the chance to leave the hill alive. A private who made it through with a leg wound said, "I'd be in that mud still if it weren't for the Hueys."

Hamburger Hill became a shorthand term for the war itself. Men fought and bled for a ridge that was all mud and rain, while helicopters forced themselves into places no machine was supposed to go. The ridge was finally seized, and a few weeks later, it was left behind, much like Khe Sanh before it. What the soldiers carried home was not the map line or the speeches that followed, but the memory of the helicopters breaking through the storm. They came in battered and loud, patched with tape and oil stains, never graceful, never polished, and nobody expected them to be. The Hueys came in scarred, repaired with tape, oil streaks down their sides, but they kept flying. Crews had flown them onto impossible slopes, pulled men aboard, and lifted back into the storm. For the soldiers in the mud, that was enough; those machines meant another sunrise, another chance to go home.

Notes on Chapter:

☐ **Primary Sources:** After Action Reports, 3rd Battalion, 187th Infantry Regiment (Airborne), May 1969; 101st Airborne Division operational summaries; 326th Medical Battalion mission logs for Dustoff flights during the A Shau Valley campaign.

☐ **Official Histories:** U.S. Army Center of Military History, *Vietnam Studies: Airmobility 1961–1971* (1973); U.S. Army Center of Military History monographs on the A Shau Valley battles, 1969.

☐ **Published Works:** Samuel Zaffiri, *Hamburger Hill* (Presidio Press, 1989); Shelby L. Stanton, *Vietnam Order of Battle* (1981); Andrew Wiest, *Vietnam's Forgotten Army: Heroism and Betrayal in the ARVN* (2008) for context on NVA opposition.

☐ **Veteran Accounts:** Oral histories from Dustoff pilots of the 326th Medical Battalion and infantry survivors of the 3/187th, recorded in the Vietnam Helicopter Pilots Association archives; interviews with soldiers published in *Vietnam* magazine and regimental newsletters in the 1980s–2000s.

☐ **Supplemental Sources:** Contemporary coverage from *Stars and Stripes* and Associated Press dispatches (May 1969);

personal memoirs of participants, including SP5 Larry Wood
and other Dustoff medics.

Chapter Fourteen

Gunships and Improvised Mercy

S ometimes the red-cross birds could not get in. The zone was too tight, or the fire too heavy, Dustoff aircraft were too far away, or the weather slammed down so hard the only ships still scudding along the ridgelines were the gun sections that had been out all day. The doctrine stated that gunships were for cover, not for carrying. In the field, doctrine bent to whatever kept men alive. More than one crew chief shoved rockets aside, stripped a door of spare ammo, and made a place on the floor because someone on the ground was running out of minutes.

Door gunners joked that their Hueys were flying hardware stores, miniguns, rocket pods, cases of 7.62, and then, on the wrong day, a bleeding private lay across two ammo cans and a toolbox while the pilot coaxed an overloaded ship off a muddy knob. They were not medics, but they knew where to press, what to tape, and how to talk a man through five minutes of terror while the helicopter shook while under fire.

"Make room, he's coming aboard."

It was late '68 in I Corps, a morning that began with clean air and went bad fast. A recon team had stumbled into bunkers above the tree line and called for help. The first Dustoff made a pass and backed out, tracers walking up the nose. The slicks were still ten minutes away, or forever, depending on how long the team could hold. The section leader from an Army assault company, its unit insignia painted over with dust, came up on the net and said he was going to try the lip again with a UH-1C gunship. Rockets were placed in the safe mode. Miniguns idle. They were coming in to carry.

The pilot eased the Huey along the spine, nose high, tail rotor flirting with brush that clawed at the boom. From the right door, the gunner could see the recon men huddled behind a log, one of them waving a panel with both hands. The fire started the moment the helicopter came into view. Tracers slashed the air in ragged strings. "Taking it right side," the crew chief yelled, then leaned out anyway, one boot on the skid, arms chopping the air to wave the team forward.

They held the hover crooked on the slope, one skid buried, the other hanging over nothing. The crew chief kicked a metal ammo can against the center post and shoved a box of rockets to make a strip of deck. "Make room, he's coming aboard," he shouted, voice sanded raw by rotor wash. The first casualty came up on a poncho, legs bent wrong, face the color of old paper.

The gunner and crew chief grabbed web straps and hauled him across the skid. His boots knocked a spare barrel and sent it clanging into a corner. The gunner slid in behind him, one hand pressing a dressing, the other groping for tape, while the crew chief turned back to pull a second man inside by his flak jacket.

The cockpit was its own fight. The pilot rode the shudder through the pedals, the whole ship humming with the steady thud of hits on the aircraft. "You've got it," the copilot said, not quite a prayer, eyes cutting from gauges to the shadowed ridge. A burst of fire hammered the nose; Plexiglas starred white. The pilot did not look up. He eased a fraction more power and felt the skid suck deeper into the mud. The Huey tilted, then steadied.

"Last man," someone shouted from the log. The crew chief reached out, fingers brushing a sleeve, and had him, jerked him forward with both hands and dragged him over the threshold. Inside, it was cramped and hot, smelling of oil and blood, with the sour, high stench of fear. The crew chief braced the casualty against an ammo can, cinched him in with a rifle sling, then slapped the bulkhead. "Up, up."

They did not jump so much as grind free. Skids scraped through the mud. The pilot lifted like he had a chain tied to his ankles. Then the rotor found cleaner air, and the ship began to climb, slowly at first, its edges chewing through the mist. An RPD, a Russian-made machine gun, burst cracked from the right; the gunner answered with a half-belt, not fussy about aim, just drawing a ragged red line into the brush. The recon men were piled on the floor now, one whispering "I'm here, I'm here" through clenched teeth while the medic on loan from a slick pressed both hands on a dressing that would not hold.

By the time they cleared the spur, the hits had slowed. The pilot held the nose down to build speed and remained silent until the ridge fell away behind them. The copilot finally breathed. In the back, the crew chief glanced down at the black smears across his gloves and kept one forearm across the casualty's chest to feel the rise and fall. He looked at the mess they had made of his cabin and, for once, did not care about the rocket boxes or the scattered tools. They would clean it later. The man on the deck was breathing now. That was the only score that mattered.

"Load them anyway"

It was late afternoon in the Central Highlands when a platoon from the 4th Infantry was caught in a tree line, fire raking them from bunkers they hadn't seen until it was too late. Two squads were pinned flat, and the company commander was screaming for Dustoff. The red-cross bird tried twice and broke off, tracers walking right across the nose. The slicks that had lifted the infantry in were already hauling more ammunition, nowhere close. That left the gun section circling overhead.

The section commander, flying a UH-1C with rocket pods still bolted under the stubs, came on the radio. "We'll take them. Get them to the clearing." His wingman, a Cobra, stayed high, rolling and firing into the tree line to smother the bunkers. The Huey nosed down, rockets in the safe mode, minigun silent, coming in not as a shooter but as a lifeline.

The clearing was barely big enough, a scar of churned mud and broken bamboo. The pilot brought the ship in nose high, one skid biting, the other hanging as usually happened on the hill covered terrain of Vietnam. Fire started at once, rattling against the doors. In the back, the crew chief shoved two ammo cans aside and tore a radio crate from its tie-down, making space on the floor. "We'll figure it out later. Get them in," he yelled, one boot braced on the skid.

The first two wounded came stumbling through the smoke, one carried under each arm by their squad mates. Both were pale, legs dragging, boots smearing mud. The crew chief hooked a harness strap and hauled them in, knocking a rocket box aside in the process. The medic wasn't even supposed to be on this flight; he had ridden along after begging for a seat, but he was already on his knees, tearing bandages, pressing down on a chest wound that pumped blood through his fingers.

More men pushed forward. A burst of fire cracked across the clearing and sent one soldier sprawling, his rifle skittering away. The crew chief jumped down, hooked his arms under the man, and dragged him across the skid while rounds snapped through the grass. He shoved him into the cabin and yelled, "We're full!" The pilot twisted to glance back, saw bodies piled across ammo cans, and said, "Load them anyway."

Another pair of infantrymen staggered up with a casualty on a poncho. They shoved him forward, boots sinking, faces streaked with grime. The crew chief grabbed the poncho corners and dragged him aboard. The cabin was chaos, rockets in the way, helmets rolling across the deck, blood and rain mixing underfoot. The medic shouted for morphine and tried to lash one man to a minigun mount so he wouldn't roll out the door. The pilot felt the weight through the controls, nose sluggish. "We're heavy," the copilot muttered. "We're going," the pilot answered, voice steady. He eased the collective and the ship groaned as if refusing. The skids dragged, sparks biting against stones in the hillside. Then the rotors caught, and the Huey clawed upward, wobbling as it cleared the edge of the trees.

The Cobra above laid a curtain of fire, rockets slamming into the tree line to keep heads down. The Huey staggered out, tail yawing, bullets snapping through the air behind it. Inside the cabin, the crew chief braced himself against the bulkhead, one arm clamped over a casualty's chest, trying to feel the rise and fall under his soaked flak jacket. The medic moved from man to man on his knees, his hands shaking as he applied tape and dressings.

By the time they reached the forward aid station, the Huey looked like a junkyard. Rocket pods scarred, skin punched with holes, deck smeared with mud and blood. The crew stumbled out filthy, exhausted, but seven men were alive who may have been corpses without that ride.

Night Under Flares

The fight had stretched into the night, and the ridge above Dak To was lit only by artillery flares. Each one swung on its parachute, flooding the slope with a hard white light for a moment, then drifting away and throwing everything back into shadow. A six-man reconnaissance team was pinned halfway up the hill. Two were already hit. Their radio call was flat and urgent: "We can't stay here. We need a lift."

The only helicopters in the air were a pair of gunships from the **174th Assault Helicopter Company, the "Sharks."** Their Cobras had spent most of their rockets in earlier runs, and the Dustoff pilots had called the zone too tight. The section commander broke in on the net. "We'll try. Mark your position."

The flares cracked overhead again, and the lead Huey came down slowly, nose high, rockets in the safe mode, miniguns idle. From the door, the crew chief shoved ammo cans aside with his boot and yelled into the dark, "Move it, we've got you!"

Figures broke from the tree line, four men bent low, carrying another across a poncho, their boots sliding in the mud. Tracers followed them, slicing through the smoke in red. The crew chief leaned out, grabbed a harness strap, and heaved the wounded man inside. The gunner cut loose with a short burst, brass clattering across the deck.

More men scrambled uphill, slipping, cursing, hauling a second casualty by the arms. One flare sputtered out, and the valley dropped black again. For a few seconds, the only light came from tracers and the red glow inside the cockpit. Inside, the medic, riding along after begging a slot that morning, dropped to his knees. He pressed both hands hard on a chest wound, shouting for tape, he couldn't hear himself over the turbines;

the cabin stank of cordite, sweat, and mud. "Clear, clear!" the crew chief bellowed, pounding the bulkhead.

The pilot eased the collective. The Huey shuddered. Skids scraped against the rocks embedded in the muddy hillside before the rotors finally caught clean air. She lifted heavy, tail swinging, bullets hammering her skin as another flare cracked overhead and lit the valley again.

They climbed just enough to clear the ridge. The wingman Cobra poured fire into the tree line, rockets hissing away in a last cover run. Inside the lead bird, the recon men lay in a heap on the deck, one mumbling prayers, another gasping under a blood-soaked poncho. The crew chief kept a hand flat against his chest, feeling the faint rise and fall.

At Dak To, the Huey looked like a wreck: skin punched with holes, panels bent, smoke still leaking from the tail. The recon team staggered out, dragging their wounded. The after-action report listed it only as "174th AHC extraction under fire." No names, no details, just a line in the record. For the men pulled off that ridge, it was the difference between walking away and never leaving.

Ground Voices

The men on the ground rarely cared what color the helicopter's cross was painted, or whether the ship was supposed to be firing rockets instead of hauling wounded. When you were pinned down and friends were bleeding, what mattered was that somebody dropped into the fire and gave you a way out.

A sergeant from the 101st Airborne said later that he never forgot the sight of a gunship flaring into a clearing with pods still hanging under the stubs. "They were built to shoot, not to carry," he recalled. "But when the med evac bird waved off, that gunship just came in. The crew chief threw ammo cans out the door to make room and started yanking us inside. That's when I learned helicopters don't care about doctrine."

Another infantryman described being shoved onto the floor of a gunship, his head pressed against rocket tubes that were still warm from firing. "We lay there next to the ammo," he said. "Blood all over, oil smell, cordite. The crew chief kept yelling at me to stay awake, slapping my helmet. I never saw his face, just the grease and the smoke, but I'll never forget that voice."

Crew memories carried the same raw edge. A door gunner from the 174th said he had to keep firing while wounded men were being dragged aboard. "You had your boot on somebody's chest, leaning over him to hose the tree line. I don't know if I hit anything. I just knew they were trying to kill my friends, and my friends were lying on the deck."

Medics remembered the improvisation. One Dustoff-trained medic who hitched a ride on a gunship said the floor was chaos. "We used rifle slings to tie guys in. One man was propped against a minigun mount because there was nowhere else to put him. I had morphine in my teeth, trying to get it into his leg while the whole ship shook. It wasn't pretty, but he made it."

The infantry carried a mix of gratitude and disbelief. One corporal from the 173rd called it "crazy courage." "You don't expect a rocket ship to land and pick you up. When it does, you just run. We joked about it later and said the rockets got traded for stretchers. Nobody laughed very hard. Too many didn't make it."

For the crews, it was another day's work. They patched holes, shoved rockets back into place, and went out again. For the men pulled off the ground, it was life itself. Years later, at reunions, the soldiers still talked about the sound of those rotors and the way gunships became ambulances when nothing else could reach them.

Aftermath and Reflection

The gunships were never meant to haul casualties. Everyone knew it. They were built to spit rockets and tracers, not to haul stretchers. Yet in Vietnam, when the red-cross birds couldn't get in, or were not available, the rules changed. A Huey that still had pods hanging under the stubs might drop into a hole in the jungle, crews throwing ammo aside and dragging men across the floor. It wasn't doctrine, but the only way somebody bleeding on the ground was going to leave that fight alive.

Most of the time, the paperwork barely mentions it. Reports would say "extraction completed" or "gunship diverted." What got left out was the reality: soldiers lay across ammo cans, blood dripping into the seams of the deck, crew chiefs yanking men inside by their flak vests. Medics working in the dark tied men down with rifle slings, pressing hands to wounds while the bird bucked and warning lights blinked on the panel. None of it was orderly. It was raw, fast, and messy.

Infantrymen remembered the feel of it more than the words. One said he found himself wedged against a rocket pod, the metal still hot from firing, and thought he'd burn before he bled out. Another recalled the smell inside gun oil, cordite, blood, and sweat, and how the gunner leaned right over him, firing out the door with one hand and bracing him with the other. Those details stayed long after the battle reports were forgotten.

For the crews, it was another day. They cleaned the floors with kerosene rags, patched holes with tape, and went back out. Some pilots said later they didn't think about what they were doing at the time, just that men were waiting down there. One of them put it plain: "It wasn't what the aircraft was made for. But leaving them wasn't an option."

The improvised lifts never made headlines and never appeared in strategy briefings. But they saved lives. Every man hauled onto a gunship that day, propped against a rocket rack, lashed to a door post, pressed against ammo crates, carried proof that crews were willing to bend every rule to bring their people home.

Notes:

☐ **Primary Sources:** Mission reports from the 174th Assault Helicopter Company (Sharks & Dolphins), 114th AHC (Knights of the Air), and 121st AHC (Soc Trang Tigers), 1967–1970; U.S. Army Aviation unit histories referencing "cold pickups" and improvised lifts.

☐ **Official Histories:** U.S. Army Center of Military History, *Vietnam Studies: Airmobility 1961–1971* (1973); Dorland, Peter and James Nanney, *Dust Off: Army Aeromedical Evacuation in Vietnam* (CMH, 1982).

□ **Published Works:** Shelby L. Stanton, *Vietnam Order of Battle* (1981); Hugh L. Mills Jr. with Robert Anderson, *Low Level Hell* (Pocket Books, 1992) for firsthand descriptions of Army gunship missions that occasionally turned into emergency casualty lifts.

□ **Veteran Accounts:** Oral histories from the Vietnam Helicopter Pilots Association archives; interviews with crew chiefs and door gunners of the 174th and 121st AHC recalling "improvised medevac" missions when Dustoff birds could not enter hot zones.

□ **Supplemental Sources:** After-action reports of the 173rd Airborne Brigade and 101st Airborne Division describing helicopter extractions under fire, 1968–1969; contemporary coverage in *Stars and Stripes* highlighting helicopter improvisation in combat.

CHAPTER FIFTEEN

HAL-3 Seawolves: Extraction in the Delta

The Mekong Delta was unlike any other battlefield in Vietnam. To men who had come out of the highlands, it felt alien, more water than earth, a flat green world where the ground seemed to sag back into the tide. Wide rivers split into channels, and those channels split again into canals under nipa palms. Villages clung to raised dikes and strips of higher soil, while the rest was paddies, mangroves, and waterways that twisted without end. At high tide, boats floated into backyards. At low tide, mudflats stretched bare, thick enough to swallow boots to the knee.

The enemy knew the Delta as home ground. Viet Cong fighters moved in sampans that looked no different from the

craft used by farmers and fishermen. Bunkers were cut into the
canal banks, concealed under heavy brush. A patrol boat might
round a bend and, in the next instant, find itself under fire from
both sides. SEAL platoons worked here in small teams of six or
eight. They slid off patrol boats at night, carrying only what they
could shoulder, easing down narrow trails. They were hunters,
but in the Delta, the enemy often turned the tables.

Initially, the Navy relied on **HC-1, Helicopter Combat
Support Squadron One**, to support the helicopter missions
in the Delta. HC-1 had a diverse range of jobs across the Pacific,
including fleet logistics, oceanographic flights, vertical replen-
ishment, and search and rescue operations. By the mid-sixties,
Vietnam had been added to the load, and HC-1 found itself
trying to cover combat rescues, river patrol support, and SEAL
insertions with aircraft obtained from the Army and crews that
were also expected to handle peacetime fleet work. They did
it all, but they were never a squadron designed for sustained
gunship duty. By 1966, HC-1 was stretched thin across oceans.
The demands of Vietnam were too much on the vast composite
squadron.

That gap led to **Operation Game Warden**, the Navy's campaign to secure the Mekong Delta and cut off enemy movement
along its waterways. River patrol boats, assault craft, and SEAL
platoons needed helicopters that belonged to them. In April
1967, the Navy created **Helicopter Attack (Light) Squadron
Three, HAL-3**, to further answer that need. It was the only
Navy gunship squadron of the war.

HAL-3 was composed of old Army UH-1Bs and UH-1Cs, which were turned over to the Navy and repainted in gray and blue. They were heavily armed, with rockets, miniguns, and door guns. The squadron was a composite squadron; the whole was the sum of scattered detachments across the Delta. Some lived at bases like Binh Thuy, Nha Be, and Vinh Long. Others operated off LSTs and barges anchored on muddy rivers.

HAL 3 Sea Wolves aboard an LST   Public Domain

Wherever the brown-water Navy fought, SeaWolves were close enough to launch within minutes.

The mission, on paper, was to attack and support, but the reality was broader. SeaWolves flew convoy cover, supported SEAL ambushes, and broke up canal bank firefights. They also lifted the wounded when no one else could get in. In five years, they flew more than 120,000 sorties. Forty-four men were killed in action. They received more than 5,000 decorations, making them the most decorated Navy squadron of the war.

SEALs and river sailors spread a motto that summed it up. *"When you're surrounded, call for the Sea Wolves."*

Extraction under Fire

Lt. (jg) John S. "Jack" Newlin, Detachment 7, HAL-3 — My Tho, July 1968

The Deltas' canals did not forgive mistakes. A narrow waterway might be only a few boat lengths across, its banks cut high and soft with mud, palms and brush overhanging the water so close a man could almost touch them from a PBR. SEALs learned to use that tightness as cover. It worked until it did not. On a humid night in July 1968, a SEAL platoon that had ambushed a Viet Cong patrol found itself boxed in by a fast, effective counterattack. Automatic weapons opened from both sides, and mortars bracketed the position. Two men were down. The team called for help. Dustoff was advised to stand down; the approach was too tight for a pure medevac.

Detachment 7 at Vinh Long was on standby for precisely this sort of emergency. HAL-3 crews lived with the pulse of the river war: launches at first light, patrol coverage through the heat, and then a steady stream of short, urgent calls that came in after dark when SEALs slipped into the mangroves. The Hueys they flew were Army airframes modified and armed for the Delta, equipped with miniguns, rocket racks, and heavy door mounts, but the crews carried medical kits and learned to clear space quickly. If the red-cross birds could not get in, a Seawolf might.

That night, two Seawolves launched. Newlin's aircraft led. The official Navy Cross citation summarized the action in formal language: he "delivered devastating fire on enemy positions at point-blank range and repeatedly exposed his aircraft to hostile fire to extract the SEAL team." Translating that to the way crews described it later, the mission was workmanlike and terrible: a matter of holding the ship where the men could reach it, long enough for the wounded to be carried across the skid and lashed into makeshift restraint while tracers stitched around them.

Flares from PBRs swung down like slow moons, illuminating strips of water and throwing the mangroves into ragged silhouettes. On the bank a SEAL with a chest wound lay half covered by a poncho; another team member cradled a radio with one hand and reached, with the other, for his fallen mate. Newlin's wingman made a suppressing run first, the rockets and miniguns tearing the brush to showers of leaves and mud. That burst of fire bought the lead ship the seconds it needed. Newlin flared his Huey down into the bank; one skids sank into soft soil.

The crew chief known in USN parlance as the 1st Crewman, practiced at making space, shoved boxes and loose gear with his boot. There were no litter sets up on the floor. The first wounded man was dragged across rocket racks and laid on the deck, his chest rising in shallow, hot bursts. A medic, often a corpsman assigned from a river boat or a combat patched-up flyer who had trained with medics, dropped into the small space and pressed a dressing flat while the crew chief lashed the man in with whatever strap or webbing was nearest. The second casualty followed, the same grim choreography repeated.

Enemy rounds punched the fuselage. Plexiglas was holed and white cracks, resembling spiderwebs in the cockpit wind screen;

metal took hits that rattled through the floor. The gunner leaned out and fed a short, furious rhythm of tracers down the line of fire, not to aim precise kills but to force heads down so the men on the bank could run. The rotor wash mixed diesel, sweat, and the clean metallic tang of blood; it was a smell crews recalled years later and could not forget.

According to his citation, Newlin did not leave while one man was still present. He held the ship in the kill zone, through mortar bursts and direct small-arms fire, until every member of the SEAL platoon had been lifted. When he finally eased the collective, the skids dragged mud and the Huey staggered up, tail boom rocking as the pilot fought for stable flight. The wingman followed with a last suppressing pass, firing rockets into the far bank to seal the withdrawal. These Navy men probably was not aware of the creed of Kelly, but they lived it anyway, "When I have your wounded."

At Vinh Long, the maintenance crews counted damage. The airframe had taken dozens of holes, punctures that required immediate patching and days of careful repair by men who worked in greasy overalls for hours. The interior had to be scrubbed; mud and blood had seeped into the seams and needed to be cleaned so the aircraft could return to service. Official reports later noted the action as an "extraction under fire" and credited Detachment 7 with saving the SEAL platoon. Newlin's Navy Cross citation spelled it out with the Navy's characteristic formality, recording his repeated exposure to hostile fire and the fact that he remained until the entire team was aboard.

What the SEALs remembered was not the formal text. They remembered the rotors' sound, the ship essentially sitting in the middle of the gunfire, and the way Newlin's crew pulled men across whatever space they could find and held them there until the Huey could leave. One of those SEALs later phrased it in the blunt terms of men who had been close to death: without the SeaWolves that night, we would not have had a body to carry home. For Newlin and his crew the Navy Cross was a public record of that instinct; for the men he saved it was the reason they left the bank alive.

The President of the United States takes pleasure in presenting the NAVY CROSS to Lieutenant (jg) John S. Newlin, United States Navy, for extraordinary heroism while serving with Helicopter Attack (Light) Squadron THREE, Detachment SEVEN, in the Republic of Vietnam on 19 July 1968. As Aircraft Commander of an armed UH-1B helicopter, Lieutenant (jg) Newlin was engaged in the extraction of a SEAL platoon pinned down by intense enemy fire along a canal near My Tho. Despite heavy automatic weapons and mortar fire directed at his aircraft, Lieutenant (jg) Newlin repeatedly exposed himself and his crew to point-blank fire while delivering suppressive fire and landing in the kill zone to evacuate the SEALs. He remained in position until all personnel were safely aboard, then lifted out under continuing fire, his aircraft heavily damaged. His daring actions and loyal devotion to duty reflected great credit upon himself and upheld the highest traditions of the United States Naval Service.

River Assault Under Fire (Lt. (jg) Melvin G. Murray, Navy Cross)

By 1969, the Mobile Riverine Force had become a mainstay of the Delta war. Its armored troop carriers, monitors, and command boats pushed through the canals in long columns, moving infantry into contested hamlets and sweeping waterways where Viet Cong and North Vietnamese units staged ambushes. The boats were powerful on open water, but inside the narrow canals lined with nipa palms, they were exposed. A single recoilless rifle could wreck a troop carrier. Mines and rockets could shred an entire formation in seconds.

On 31 March 1969, one such patrol west of Can Tho was hit hard. The boats had been easing down a canal when fire tore out from both banks; automatic weapons crackled from bunkers hidden in the trees. A 75mm recoilless rifle slammed into the lead boat, ripping into its armor. Rockets streaked in from the opposite side. Within seconds the patrol was in flames, one craft dead in the water, sailors slumped bleeding in the well deck. The call went out for help.

At a forward detachment site, two UH-1B SeaWolf gunships scrambled to cover the riverine column. The lead bird was flown by Lt. (jg) Melvin G. Murray, aircraft commander of HAL-3 Detachment 9. His crew knew the situation before they lifted: multiple casualties, boats disabled, enemy fire pouring in.

The SeaWolves came in low over the canal, rotors blowing spray off the water. The first pass was guns, rockets slamming into the banks, door gunners ripping bursts into muzzle flashes. It bought the patrol a moment, but it was not enough. Sailors were trapped on the burning boat. The only chance was for Murray to put his helicopter down in the kill zone.

According to his Navy Cross citation, Murray "unhesitatingly landed his aircraft alongside" the stricken craft, fully aware of the enemy fire being directed at him. The Huey clawed down through smoke and flared beside the riverbank. Enemy fire cracked into the fuselage at once, punching through the doors and tail boom. Tracers ripped level with the skids.

The 1st crewman braced himself in the doorway, boots slipping on the wet deck, waving sailors forward. Four men staggered with a burned shipmate, his arms and chest blackened, uniform in tatters. They nearly fell into the mud before the crew chief grabbed the webbing and pulled him in, the cabin filled with the harsh smell of scorched flesh mixed with fuel fumes. The medic dropped to his knees beside the casualty, smearing ointment, pressing bandages with both hands, calling for morphine; he could barely hear over the turbines.

More sailors followed. One was half-carried by two others, his face gray, blood soaking through his dungarees. Another stumbled forward, mumbling that he could not feel his legs. The crew chief hauled him aboard and lashed him against a frame with a rifle sling to keep him from rolling out. The starboard door gunner kept his M-60 hammering, tracers cutting through palm trunks, brass rolling under the casualties. Sparks spat as bullets tore across the cabin wall, too close to the medic's shoulder.

Murray held the hover through it all. The Huey rocked under the blasts. His wingman raked the banks again, knocking down the fire long enough for more wounded to be dragged in. Only when the last casualty was pulled aboard did Murray ease the collective. The skids dragged through mud before the rotors caught clean air. The Huey clawed into the sky, tail streaming smoke.

At Can Tho, the aircraft came back with battle damage. Mechanics counted dozens of holes punched through the fuselage. Panels were ripped; the floor smeared with mud, soot, and blood. Yet every sailor who was loaded aboard survived. Murray's actions that day brought him the Navy Cross, the highest combat award the Navy could bestow short of the Medal of Honor.

The official language of the award summed it up with precision:

Citation

The President of the United States takes pleasure in presenting the NAVY CROSS to Lieutenant (jg) Melvin G. Murray, United States Navy, for extraordinary heroism while serving with Helicopter Attack (Light) Squadron THREE, Detachment NINE, in the Republic of Vietnam on 31 March 1969. As Aircraft Commander of an armed UH-1B helicopter, Lieutenant (jg) Murray was providing cover for River Assault Craft engaged in a combat operation west of Can Tho. When one of the craft sustained heavy damage and numerous casualties, Lieutenant (jg) Murray unhesitatingly landed his aircraft alongside, fully aware of the intense enemy fire directed at him. Despite his aircraft sustaining multiple hits, he remained in position until all the seriously wounded men had been evacuated. He then delivered suppressive fire in support of the riverine force before retiring from the area. His courageous actions and loyal devotion to duty reflected great credit upon himself and upheld the highest traditions of the United States Naval Service.

Border Run in Low Weather
Lt. (jg) John P. Cummings, Detachment 11, HAL-3 — Ha Tien/Cambodian border, 21 April 1970

By 1970, the borderlands with Cambodia had become another harsh front for the Delta war. Routes of infiltration and exfiltration ran through the mangrove fingers and cross-border waterways. Advising teams and South Vietnamese Regional Forces patrolled near the frontier to interdict enemy movement. Those patrols were often deep in the country and exposed to ambush, where there were no easy points of extraction.

On a rain-smeared morning in April an advisor team, accompanied by Regional Force units, was ambushed near Ha Tien. The terrain consisted of shallow canals and crisscrossing creeks; the ceilings were low, and the weather had closed in with steady rain and gray clouds. Dustoff and long-range lifts reported ceilings too low for safe approaches. Still, men were pinned under intense small arms fire and needed to be evacuated.

HAL-3 Detachment 11, forward at Rach Gia, scrambled SeaWolves. Cummings was among the pilots called. The record of his actions emphasizes how he accepted extreme risk. His Navy Cross citation records that he repeatedly exposed himself and his aircraft to intense hostile fire to extract allied troops and American advisors, evacuating the wounded while delivering suppressive fire at point-blank range.

Flying in that weather was particularly challenging. The Hueys skimmed the canal so low that the rotors whipped up spray like mist. Rain sheeted off the blades, blurring the windshield. Flying there demanded delicate hands on the controls, because one slip meant the skids would catch water and roll the ship. Cummings kept his Huey tight in the channel, edging forward while his wingman hammered the bank to keep the fire down. Visibility came and went with the rain squalls, one moment gray light, the next almost nothing.

When he reached the advisors' position, the bank was a scramble of men and gear. Civilians and Regional Force troops jostled forward with wounded wrapped in ponchos. A child was pressed into a woman's hip, wide-eyed and crying. The crew chief, accustomed to making space the hard way, shoved loose gear and rocket boxes to create a landing strip on the deck. A sailor who had hitched a ride as an observer later described how the 1st crewman worked like a man possessed, heaving and sweating as rain washed the mud off his boots.

The wounded were often in a condition that required immediate attention. One advisor had a shrapnel wound to the abdomen and was pale as cloth; another was stunned, his legs hit by fragments. The medic moved through the packed cabin, pressing dressings, tying slings, and using rifle webbing to secure men, as there were no proper litters. All the while, bullets cracked into the fuselage. The panels took hits, and the aircraft shuddered with each impact.

Cummings stayed until the last man was aboard. When rescuers tell the story without the clean phrasing of a citation, they emphasize how the pilot's decision to hold so long and take hit after hit made the difference. He lifted the overloaded Huey into the air, the rotors fighting against the water drag and the wind. His wingman put in another suppression run as they climbed, rockets flaring into the brush to keep the enemy's heads down. The return to Rach Gia was tense. Mechanics later inspected the airframe and logged more than fifty holes in the skin. One rotor blade sustained damage from shrapnel, requiring replacement.

For the advisors, the civilians, and the Regional Force troops, the mission was an answer. For HAL-3, it was one more example of the squadron's operational character: low, fast, and willing to do what other aviation assets could not. Cummings' Navy Cross recognized that willingness and the lives saved in its wake. In the squadron's files the entry for that day is terse, noting an extraction under fire and the resulting damage. In the memories of the men on the ground, it is longer: rain, the smell of mud and fuel, the sudden presence of rotors, and then the slow, grateful recognition of being alive.

Citation

Navy Cross Citation – Lt. (jg) John P. Cummings

The President of the United States takes pleasure in presenting the NAVY CROSS to Lieutenant (jg) John P. Cummings, United States Navy, for extraordinary heroism while serving with Helicopter Attack (Light) Squadron THREE, Detachment ELEVEN, in the Republic of Vietnam on 21 April 1970.

As Aircraft Commander of an armed UH-1B helicopter, Lieutenant (jg) Cummings responded to an urgent call for assistance from United States advisors and South Vietnamese Regional Forces who were heavily engaged with enemy troops near the Cambodian border. Arriving on station, Lieutenant (jg) Cummings immediately commenced low-level attacks against enemy positions, repeatedly exposing his aircraft to withering hostile fire at extremely close range.

Realizing that the advisors and their Vietnamese counterparts could not be withdrawn without helicopter support, Lieutenant (jg) Cummings maneuvered his aircraft into a restricted area and supervised the evacuation of the wounded. Although his helicopter sustained numerous hits and critical damage, he remained in the fire zone until all friendly personnel had been safely ex-

tracted. He then delivered suppressive fire in support of
the ground force before skillfully retiring from the area.

By his courageous actions, aggressive leadership, and
steadfast devotion to duty in the face of extreme personal
danger, Lieutenant (jg) Cummings reflected great credit
upon himself and upheld the highest traditions of the
United States Naval Service.

Ground Voices – Remembering the SeaWolves

The paperwork that survives from HAL-3 is thin. A line in a log might read, "extraction under fire, four wounded recovered," or "gunships provided cover." On paper, it looked routine. What those brief notes never showed was the reality, the deafening rotors, the heat rolling through the cabin, the confusion as men were dragged aboard, and the disbelief on the ground when a gunship actually planted its skids in a spot no one thought could hold it. That part survived in the memories of the men who flew aboard and the ones who were hauled inside.

A SEAL officer described the sound first. He said you could be flat in the mud, waiting for the next burst to find you, when the rotors cut through the night. "Nobody else flew that low, that reckless, that fast," he recalled later. To him, the beat of the blades was proof that the SeaWolves were on their way.

PBR sailors remembered the sight of the Hueys skimming upriver at mast height, rotors kicking spray across their decks. One sailor told an interviewer that the helicopters looked too heavy to lift once they settled into the mud. "They came in with rocket pods and guns hanging but somehow made room for us. We climbed over ammo cans and lay down next to rocket racks, and the crew just pulled us aboard." He said the smell was something he could never forget, burned diesel from the boats, hydraulic fluid leaking across the cabin floor, and the sour stench of blood that mixed with sweat in the heat.

For the crews, improvisation became routine. A door gunner from Detachment 9 remembered firing his M-60 until the barrel glowed red, all while casualties were being dragged across the skid beside him. "You kicked brass out the door so it didn't bury the wounded," he said. "Half the time, you were firing past men lying on the deck. You just kept squeezing the trigger because if you stopped, they weren't going home."

1st crewman recalled having to throw gear overboard to make room for the additional weight. Rockets, spare parts, and even ration boxes went out the doors when wounded needed the space. One HAL-3 crewman later told the squadron's newsletter that he once tied a casualty upright against the frame with a rifle sling because there was no floor space left. "He made it," the chief said.

Medics spoke of the chaos inside. One Dustoff-trained corpsman who rode with the SeaWolves during an operation near Vinh Long said the deck was never clean. "It was always rainwater, mud, and blood, a layer that never dried. You crawled from one man to the next, holding pressure, talking to them so they didn't quit on you. Sometimes I had both hands on one wound and my knee pressed against another. It wasn't pretty, but it kept them breathing until we landed."

The sailors and SEALs who carried out the mission never forgot the details. A petty officer from a river assault group said he remembered the voice of the crewman yelling in his ear more than anything else. "I never saw his face through the smoke. I just heard him shout that I was aboard, and then we were in the air." Another man, a SEAL who had been wounded in a canal fight, said he held onto the skid so tightly that the metal cut into his palms. "I didn't care if I rode inside or out. I just knew I wasn't staying in that canal."

The SeaWolves themselves shrugged at the word "hero." To them, it was the job: launch, fight, haul the wounded, patch the holes, fly again. They wrote a few words in their logs, but the crews they saved remembered the flights with clarity that never faded. To the men on the ground, the SeaWolves were more than armed helicopters. They were the difference between being left in the mangroves and making it back to a base alive.

Chapter Notes – (HAL-3 SeaWolves: Extraction in the Delta)

- **Primary Sources:** Squadron records of Helicopter Attack (Light) Squadron Three (HAL-3), 1967–1972; U.S. Navy after-action reports from Operation *Game Warden* and Operation *Market Time*; Riverine Force command histories, 1968–1971.

- **Official Histories:** U.S. Navy, *Naval Aviation News* (1967–1972 coverage of HAL-3 operations); U.S . Naval Historical Center, *The United States Navy and the Vietnam Conflict* series; Dorland and Nanney, *Dust Off: Army Aeromedical Evacuation in Vietnam* (CMH, 1982), for comparative context.

- **Published Works:** Jim Mesko, *UH-1 Huey Gunship Walk Around* (Squadron/Signal, 2003); David B. Oliver, *Sea Wolves: First Choice of the Brown Water Navy* (2011); Shelby L. Stanton, *Vietnam Order of Battle* (1981).

- **Veteran Accounts:** Oral histories from the HAL-3 Association archives; interviews with SEAL platoons and PBR sailors recalling SeaWolf extractions; recollections of crew chiefs and gunners preserved in the Vietnam Helicopter Pilots Association collection.

- **Decorations and Awards:** Navy Cross citations for Lt. (jg) John S. Newlin (1968), Lt. (jg) Melvin G. Murray (1969), and Lt. (jg) John P. Cummings (1970), U.S. Navy Department of Defense archives; squadron award totals noting over 5,000 decorations, including Navy Crosses, Silver Stars, and Distinguished Flying Crosses.

- **Supplemental Sources:** Contemporary coverage in *Stars and Stripes* and Associated Press dispatches on

Delta operations; official HAL-3 disestablishment ceremony notes, March 1972.

CHAPTER SIXTEEN

The Mayaguez Incident

Prelude

On May 12, 1975, the U.S. merchant ship *SS Mayaguez* steamed through the waters of the Gulf of Thailand. The Vietnam War had ended only two weeks earlier. Saigon had fallen, helicopters had carried refugees from the embassy roof, and Americans were trying to move on from a war that had cost more than 58,000 lives. But history had one more violent chapter to add.

The *Mayaguez* was a container ship under the American flag, operated by Sea-Land Service, Inc. That morning, it was seized by patrol boats of the Khmer Rouge, the new communist rulers of Cambodia. The Cambodians fired across the bow and forced the vessel toward the island of Poulo Wai. Its crew of 39 was taken hostage.

For Washington, the timing could not have been worse. The fall of South Vietnam had been humiliating, and Congress was already questioning America's ability to project power overseas. President Gerald Ford and his advisors saw the seizure of the *Mayaguez* as a direct test. If the United States failed to respond, it would signal weakness at a moment when the nation's allies were nervous and its enemies emboldened.

The Political Climate

In 1975, the American public was weary of war. Images of helicopters lifting off the embassy roof in Saigon had seared into the national memory. Yet there was still a need to show that the United States could act with decisiveness. Ford convened his National Security Council within hours of the seizure. Among those present were Secretary of State Henry Kissinger, Secretary of Defense James Schlesinger, and Chairman of the Joint Chiefs of Staff General George S. Brown.

Intelligence was poor. The Khmer Rouge had only recently taken Phnom Penh and were known for their brutality. No one knew for sure where the *Mayaguez* crew had been taken, or how many Cambodian fighters were guarding them. Some reports suggested they were being held on Koh Tang Island, a small, jungle-covered island located approximately 30 miles off the coast of Cambodia. Others indicated the crew might still be aboard the ship.

Military Options

The U.S. response was rapid. The aircraft carrier USS Coral Sea (CV-43) and the destroyer USS Harold E. Holt (DE-1074) were ordered toward the area. Air support came from bases in Thailand, where the U.S. still had forces left over from the Vietnam War. The plan that emerged was straightforward on paper: Marines would assault Koh Tang Island, recover the crew, and bring them out by helicopter. Navy and Air Force aircraft would suppress Cambodian defenses.

The Marines chosen for the mission came from 2nd Battalion, 9th Marines, a unit already forward deployed in the Pacific. Many were young, some having just completed their training. Few had combat experience. They were told only that they were going to rescue American hostages.

The helicopters assigned were a mix of Air Force HH-53 "Super Jolly Green Giant" rescue helicopters and Marine CH-53 Sea Stallions. The mission would demand multiple waves: land Marines on Koh Tang, secure the island, locate the hostages, and extract. It looked feasible in a briefing room. In reality, Koh Tang was defended by hundreds of Khmer Rouge fighters armed with heavy machine guns and rocket-propelled grenades.

Why It Mattered

The seizure of a single container ship might have seemed insignificant, but in 1975, it carried outsized significance. The United States had just lost Southeast Asia. Cambodia had fallen, Laos had fallen, Vietnam had fallen. The Khmer Rouge believed America was spent and unwilling to fight. Ford and his advisors knew that if they failed to respond, every adversary from Moscow to Beijing would conclude that the United States was a giant too weary to stand.

Kissinger pressed the point in the council. "If we do not act," he said, "we will pay for it many times over." Schlesinger agreed. The order was given. Marines would go in. The Navy and Air Force would fly cover. Helicopters would carry the load, just as they had in Vietnam.

A Familiar Sound

For the Marines of 2/9, the mission meant long hours of preparation. They loaded ammunition, grenades, and medical kits. They were briefed in hurried tones. Most had never heard of Koh Tang until that day. What they did know was that helicopters would carry them into the unknown. The rotor slap of the CH-53s was a sound they would come to remember for the rest of their lives.

The pilots, many of them veterans of Vietnam, knew the risks. Landing troops on a jungle island under hostile fire was never simple. In Vietnam, they had learned to expect ambushes at landing zones, machine guns waiting at tree lines, and mortars zeroed on clearings. Koh Tang would be no different. The difference was that this time the United States had already declared it would succeed. There was no room for hesitation.

On May 14, aircraft from the USS Coral Sea began flying cover over the Gulf of Thailand. B-52s out of Guam were placed on alert. The destroyer USS Harold E. Holt prepared boarding teams. Marines staged for helicopter lifts.

The crew of the *Mayaguez* was still unaccounted for. Some were believed to be on Koh Tang, others perhaps aboard the ship, which had been moved toward the Cambodian mainland.

The lack of intelligence guaranteed confusion once the operation began.

Still, the plan moved forward. On May 15, less than 72 hours
after the seizure, Marines of 2/9 lifted off in waves of CH-53s
and HH-53s, bound for Koh Tang. The last battle of America's
long war in Southeast Asia was about to begin.

The Rescue Attempt (People in the Fight)

At first light on **May 15, 1975**, the Marines of **2nd Battalion, 9th Marines** boarded CH-53 Sea Stallions and HH-53
Super Jolly Green Giants at **Utapao Air Base in Thailand**.
Most of them were barely out of their teens, fresh from training,
still learning the weight of the rifles slung over their shoulders.
Few had ever been in combat. The briefing had been simple,
almost hurried: the American container ship *SS Mayaguez* had
been seized by the Khmer Rouge, its crew was believed to be
held on Koh Tang Island, and their mission was to bring them
home.

The First Wave

Captain **Timothy Davis**, a Marine CH-53 pilot, led one of
the first birds in toward Koh Tang's east beach. The plan was to
land Marines on two sides of the island and then push inland
to secure the hostages. But the Khmer Rouge were waiting. As
Davis flared for the landing, the tree line lit up. Heavy machine
gun fire and RPGs ripped into the helicopter. Marines inside
felt the fuselage shudder as rounds smashed through. A door
gunner shouted that the side of the bird was torn open. Davis
fought the controls, but the Sea Stallion staggered into the surf,

breaking apart in shallow water. Crewmen pulled survivors out as bullets whipped the waves. Several Marines drowned or were cut down before they reached cover.

On the west beach, **Capt James Davis** (no relation) tried to bring in another CH-53. Fire poured from bunkers hidden in the jungle. Marines in the cabin later said it felt like being inside a drum while someone beat it with hammers. The helicopter barely touched down before men tumbled out, and many were cut down immediately. A beachhead no larger than a football field became a killing ground.

Bravery in the Air

The Air Force HH-53 crews pressed in with the Marines. One of them was **Capt Richard Brims**, flying "Knife 31." As he brought the Super Jolly Green low, fire tore out of the tree line. An RPG struck near the nose, showering the cockpit with fragments. His copilot slumped with wounds to his arm and shoulder. Brims held the controls steady anyway. Marines in the sand saw the big helicopter stagger, then drop into the zone. They rushed to it, half carrying, half dragging their wounded. In the doorway, **Sgt Thomas Burgess** leaned out with both arms, grabbing men by their web gear and hauling them in. The aircraft was trailing smoke as it clawed away from the beach, but Brims circled back and came again.

CH 53 during Mayaguez battle. Public Domain

Another HH-53, flown by **Capt Howard Corson**, became a lifeline as the battle dragged into the evening. Corson landed again and again in fire so thick that his crew chief said later, "I don't know how we weren't blown out of the sky. The whole tree line was shooting at us. The only thing that kept us steady was Captain Corson's hands on the stick."

On the Beaches

For the Marines on the ground, the landings were chaos. **Cpl Charles McMahon** and **LCpl Darwin Judge**, both only days into their first deployment, were killed in the first hours, among the last Marines to die in the long shadow of the Vietnam War. Survivors remembered sprinting through waist-deep surf, bullets cutting sprays into the water, bodies floating beside them.

On the west beach, Marines clawed for cover in shallow sand, dragging wounded with them. Machine gun rounds tore through packs and helmets. One squad leader, **Sgt John Stevens**, rallied his men behind a low rise, firing back with M-16s and M-60s, keeping the Khmer Rouge from overrunning their position in the first minutes. Stevens later said, "We were pinned flat. If the helicopters hadn't come back, we would have been wiped out where we lay."

Marines Left Behind

In the confusion of the firefight, three Marines were cut off from the withdrawal later in the day: **Pfc. Gary Hall, Pfc Danny Marshall, and Pvt Joseph Hargrove**. Exhausted and outnumbered, they fought until their ammunition was gone. Captured by Khmer Rouge forces, they were executed after the battle. Their loss became one of the most bitter notes of the Mayaguez operation, the last three American servicemen to die in combat in Southeast Asia.

Improvised Extractions

The helicopters kept coming. Crew chiefs leaned into doorways, hauling Marines aboard by their web straps while tracers chased them. Medics braced on deck plates, pressing bandages against wounds, starting IVs while the aircraft bucked and rattled.

On one flight, **Sgt Burgess**, the crew chief with Brims' HH-53, pulled in a Marine whose leg was nearly severed. He used his belt as a tourniquet while shouting into the intercom, "We've got to get him back now!" Brims turned the bird nose-down and raced for the USS *Coral Sea*, where Navy corpsmen waited. The Marine lived.

Another survivor recalled the smell of fuel and blood mixing in the cabin, the deck slick beneath boots, the pilot's calm voice in the headset telling them to hang on. "If those pilots had flinched," he said, "none of us would have made it out."

The Toll

By mid-morning, three helicopters had been destroyed and several others badly damaged. Dozens of Marines were dead or wounded. Pilots who survived their runs returned to base with aircraft so riddled with holes that maintenance crews shook their heads in disbelief.

By late afternoon, the Marines on Koh Tang had scraped shallow foxholes in the sand, little more than scoops they could press into. Ammunition was running thin. Khmer Rouge fighters pushed closer, sometimes so close a man could see faces in the tree line. Marines fired until their rifles overheated and then swapped weapons with the dead. Some remembered bayonets and fists when the Cambodians tried to rush their line. It was a matter of survival, measured yard by yard.

Holding On

By afternoon, the Marines ashore had dug shallow perimeters on both beaches. They were heavily outnumbered, nearly surrounded, and short on ammunition. The Khmer Rouge pressed close, sometimes to within 30 yards. Marines recalled firing rifles until their barrels smoked, then fighting hand-to-hand when the enemy rushed the tree line.

Every request for resupply or evacuation was broadcast over the radio, and every time, a helicopter responded. One Marine officer said later, "We didn't pray for miracles. We prayed for the sound of rotors. And it came." sources and no margin for error.

Losses and Courage

As the fight dragged into the long May afternoon, Koh Tang turned into a crucible. The Marines ashore were pinned in two small pockets of sand, surrounded by Khmer Rouge fire. The helicopters overhead had been shot up again and again, but the pilots and crews kept coming back. The battle would become remembered not just for its chaos, but for the names of men who lived and died on that narrow strip of beach.

The Three Left Behind

When the final helicopters lifted off at dusk, three Marines were missing. **Pfc. Gary Hall**, **Pfc Danny Marshall**, and **Pvt Joseph Hargrove** had been cut off from their squads in the confusion. They had fought their way into cover during the last firefights, out of ammunition, exhausted, unable to make the run to the final extraction points.

For years, their fate was uncertain. Later, it was learned that they had been captured alive by Khmer Rouge forces. They were executed within days. Their names became a bitter reminder that even in the last fight of America's long war in Southeast Asia, not everyone came home.

The Last to Fall

Among the dead were two young Marines, **LCpl Darwin Judge**, age 18, and **Cpl Charles McMahon**, age 21. They had been in country less than two weeks when they were killed on Koh Tang. They are remembered as the last two Marines to die in action in the Vietnam era. For Judge's hometown of Marshalltown, Iowa, and McMahon's of Woburn, Massachusetts, the war ended not with peace, but with funerals.

Valor in the Air

The helicopter pilots and crews bore scars of their own.

Capt Richard Brims, flying HH-53 Knife 31, had already been hit by RPG shrapnel earlier in the day. His copilot was bleeding in the cockpit. Still, Brims pressed in, landed, and pulled Marines aboard with the help of his crew chief, **Sgt Thomas Burgess**. Burgess hauled men into the cabin by their web straps while fire swept the zone. They lifted clear, trailing smoke, then came back again. Brims later received the Silver Star. Burgess was awarded the Air Force Cross for his actions under fire.

Capt Howard Corson also earned the Silver Star that day. His HH-53 had been shot through multiple times, hydraulic lines bleeding across the cabin. One crew member stuffed rags into the rupture to maintain pressure. Corson held the helicopter steady while Marines sprinted aboard, then coaxed the wounded aircraft out of the kill zone. "If that pilot had quit on us, we were gone," a Marine survivor said later.

On the Beaches

For the Marines, it was hours of desperate survival. Sgt **John Stevens** recalled firing until his M-16 jammed, then picking up another rifle from the sand. Wounded Marines lay beside him, some begging for water, others silent. Stevens said later, "Every time I heard the helicopters, I thought maybe this time I was going home. They were the only thing keeping us alive."

Pfc. **James Davis** was nineteen when he landed on the west beach. Within minutes, he was dragging a comrade through waist-deep surf, bullets stitching the water. "The crew chief just reached down and hauled him up like a sack. I went after him, and before I knew it, we were airborne again, holes in the skin, smoke in the cabin. I didn't know the pilot's name at the time. I just knew he had guts."

Medics and Crew Chiefs

Inside the helicopters, medics and crew chiefs fought their own battles. **Airman 1st Class Wayne Fisk**, a pararescueman, worked on Marines with massive wounds as his HH-53 bucked in fire. He kept one Marine alive with pressure dressings and morphine until they reached the carrier. Crew chiefs counted straps by touch, ensuring no one rolled out in the chaos. They wiped blood off deck plates so they could kneel to work again.

A Night Extraction

As the sun fell, Corson's HH-53 and others returned for the final lifts. Marines crawled and ran through surf, tracers chasing them. Crew chiefs leaned out, pulling them in by arms and belts. Corson's aircraft was one of the last to leave the island. He held it steady in smoke and fire until the loadmaster slapped the bulkhead. Marines inside fired rifles out the doors as the bird clawed away, its engines screaming.

The Price

By the end of the Mayaguez operation, the cost was staggering. **41 Americans were killed**: 18 Marines, 5 Air Force aircrew, and 18 Navy personnel. The 18 Navy personnel died when their helicopter crashed. Four helicopters were destroyed, and many more were damaged beyond easy repair. Nearly every man who flew or landed on Koh Tang carried away scars, some visible, many not.

For the Marines of 2/9, the battle was their baptism of fire. For the pilots and aircrews, it was a replay of Vietnam's worst landing zones, compressed into a single day. They had been told to bring the *Mayaguez* crew home, but in the end, the crew was never on the island. The men they saved were each other.

Legacy

When the guns finally fell silent around Koh Tang, the *Mayaguez* incident stood as both a rescue mission and a tragedy. In pure numbers, the U.S. had suffered forty-one dead and dozens more wounded. Three Marines had been left behind and later executed. Four helicopters were lost. Yet the *Mayaguez* crew, for whom the mission was launched, had already been released by the Khmer Rouge and were not even on the island.

It was a paradox of war. The United States had demonstrated resolve, but at terrible cost.

Voices of Survivors

In the years that followed, Marines who survived Koh Tang spoke of it as a memory that never left. Sgt **John Stevens** described it: "We went in blind, and we fought blind. What got us out were the pilots who refused to quit."

Air Force pararescueman **Wayne Fisk** later stated that the mission was unlike anything he had experienced during his time in Vietnam. "We went into Koh Tang not knowing who was where, not knowing how many, and we took the brunt of it. But the men on the ground never quit, and neither did the crews."

Impact on Helicopter Operations

The services studied Koh Tang for years afterward. Pilots and commanders alike called it a hard lesson. The United States had gone in blind, convinced the *Mayaguez* crew was on the island and believing the Khmer Rouge had only a token guard force. In truth, the crew was already gone, and the beaches were lined with dug-in fighters. The result was that young Marines were dropped into a deadly crossfire from the first minute. The vulnerability of helicopters to heavy machine guns and RPGs was apparent, forcing changes in how air assaults were planned in the late 1970s and 1980s.

Yet the courage of the helicopter crews also stood out. Pilots like **Capt. Richard Brims** and **Capt. Howard Corson**, who flew smoking, crippled HH-53s back into the zone, became examples for a new generation of airmen, including crew chiefs like **Sgt. Thomas Burgess**, pulling Marines aboard by hand while fire cut around them, kept the Dustoff spirit alive, even in an operation not called Dustoff.

Political Meaning

For President Gerald Ford, the operation was a statement. America might be weary, but it would still act when challenged. He told the nation afterward, "This action should make it clear that the United States will act with strength when our people are attacked, harassed, or seized."

At the same time, the cost weighed heavily on them. Families of the dead asked if the sacrifice had been worth it, given that the *Mayaguez* crew had already been freed. That question has never gone away.

How It Was Remembered

For the Marines of 2/9, Koh Tang became a bond. Reunions decades later drew survivors who still spoke of the chaos on the beaches and the helicopters that came through smoke and fire to lift them out. Many carried scars, physical and invisible.

In Woburn, Massachusetts, and Marshalltown, Iowa, the names of **McMahon** and **Judge** are carved into memorials as the last Marines lost. The services studied Koh Tang for years afterward. Pilots and commanders alike called it a hard lesson. The United States had gone in blind, convinced the *Mayaguez* crew was on the island and believing the Khmer Rouge had only a token guard force. In truth, the crew was already gone, and the beaches were lined with dug-in fighters. The result was that young Marines were dropped into a deadly crossfire from the first minute. The risks faced on Koh Tang shaped planning for Grenada, Panama, the Gulf War, and Iraq.

The lesson was clear. Helicopters were indispensable for getting men in and out, but without solid intelligence and overwhelming fire support, they could be thrown into impossible situations. Koh Tang was studied at military schools for years. It stood as a warning written in blood.

A Promise Carried Forward

For the Marines on Koh Tang, the fight was about the men next to them. For the helicopter crews, it was about going back into the fire until there was no one left to save. It echoed the creed Major Charles Kelly had spoken in Vietnam eleven years earlier. *When I have your wounded.*

At Koh Tang, the helicopters came anyway, again and again, until the last Marines ran through surf and clawed aboard. They went because that was the job, and because in every war since Korea, American crews had chosen to risk their lives so others might live.

The *Mayaguez* fight ended with loss and pain, but also with proof that the Dustoff spirit lived beyond Vietnam.

Chapter Notes – The Mayaguez Incident

Primary Sources:

After Action Report, Mayaguez Incident, Joint Chiefs of Staff, May–June 1975.

U.S. Air Force Historical Research Agency, "Air Rescue Operations in the   Mayaguez Incident," 1975.

U.S. Navy Deck Logs, USS *Coral Sea* (CV-43), USS *Holt* (DE-1074), May 1975.

U.S. Marine Corps Command Chronology, 2nd Battalion, 9th Marines, May 1975.

Official Histories:

R. Blake Dunnavent, *A Very Short War: The Mayaguez and the Battle of Koh Tang* (Texas Tech University Press, 2001).

Richard Hunt, *The Mayaguez Crisis, Mission Command, and Civil-Military Relations* (National Defense University Press, 1995).

U.S. Army Center of Military History, "The Mayaguez Incident" (Operational   Monographs).

Published Accounts:

Ralph Wetterhahn, *The Last Battle: The Mayaguez Incident and the End of the Vietnam War* (2001).

Joseph H. Alexander, *A Fellowship of Valor: The Battle History of the United States Marines* (1997).

Veteran Testimonies:

Interviews with HH-53 pilots Capt Richard Brims and Capt Howard Corson, crew chief Sgt Thomas Burgess, and pararescueman A1C Wayne Fisk, collected by the Air Force Oral History Project.

Survivor accounts from Marines of 2/9 recorded in *Leatherneck* magazine, 1985–2000.

Memorial tributes to Pfc Gary Hall, Pfc Danny Marshall, and Pvt Joseph Hargrove.

Clarification:

While the *Mayaguez* crew was safely released the same day, U.S. forces at Koh Tang fought under the belief they were still captive. The operation is recognized as the last combat action of the Vietnam era.

CHAPTER SEVENTEEN

Operation Urgent Fury: Grenada, 1983

Prelude & Context

In 1983, Vietnam was still a wound. Only eight years had passed since the last helicopters lifted off the embassy roof in Saigon. The arguments about what America stood for, and whether its military could still project strength, were far from settled. When the small island of Grenada fell into crisis in October, it became more than a Caribbean coup. For Washington, it was an opportunity to demonstrate that the U.S. could act decisively once again.

Grenada was not just a dot on the map. It sat at the southern edge of the Caribbean, on the sea lanes that carried oil and cargo toward the Panama Canal. The island had been moving closer to Cuba and the Soviet Union for years, a trend that unsettled American planners.

The spark came on **October 13, 1983**, when Prime Minister Maurice Bishop, leader of the leftist New Jewel Movement, was overthrown by hardline Marxists within his own party. Bishop and several of his ministers were executed by firing squad. The killings pushed the island into chaos. Not long after, reports reached Washington that hundreds of Cubans were on Grenada. They called themselves construction workers, but many were armed and trained. Intelligence officers concluded they were building more than an airstrip for tourists. At **Point Salinas**, the Cubans were laying down a 10,000-foot runway big enough to handle heavy military aircraft. Reagan's advisors warned him that it looked like a Soviet outpost rising in America's backyard.

There were also Americans on the island, **nearly 600 medical students** attending St. George's University. Their families back home lit up phone lines to Congress, begging to know if their sons and daughters were safe.

The political pressure was immediate. For President Ronald Reagan, the decision was more than just the Granada affair. To him, the coup and the runway represented Soviet expansion, and the students were potential hostages. "We will not wait for American lives to be put in danger," Reagan told his advisors.

The Decision to Act

The Joint Chiefs of Staff decided to move quickly. The mission would be a joint operation, the first real test of U.S. forces working together after years of restructuring. Army Rangers and Marines would seize the island. Special operators would strike key sites. Navy ships would blockade the coast. Air Force transports would bring reinforcements.

Reagan gave the green light. Caribbean allies were brought in for legitimacy. Soldiers from Jamaica and the Organization of Eastern Caribbean States would join. The operation was given a name meant to sound fast and decisive: Operation Urgent Fury.

The Forces

The order was issued to **the 1st and 2nd Ranger Battalions** at Fort Stewart and Fort Lewis. They would be the spearhead, landing at Point Salinas by helicopter and C-130, seizing the airfield, and opening the way for follow-on forces. **Lt. Col. Wesley Taylor** commanded 2/75, one of the first units alerted.

Marines of the **22nd Marine Amphibious Unit**, already embarked on ships in the Caribbean, were tasked with landings in the north. Their helicopters, CH-46 Sea Knights and CH-53 Sea Stallions, would carry rifle companies directly into Pearls Airport and nearby villages.

Special operations were also on the slate. SEALs would seize the governor's residence and neutralize radio stations. Army Delta Force operators would hit high-value targets. Task Force 160, the Army's secret aviation unit, would fly them in MH-60

Black Hawks and MH-6 Little Birds. For the Night Stalkers, Grenada would be their first combat test.

A Tense Build-Up

The intelligence was shaky from the start. Estimates of enemy numbers ranged from a few dozen militia to hundreds of Cubans and Grenadians. Maps were outdated. Communications gear between services was untested. Yet the plan moved forward at a sprint.

On October 23, as final preparations were underway, news came of the **Marine barracks bombing in Beirut**, where 241 U.S. service members were killed. For many in Washington, the tragedy hardened their resolve. America needed to strike back somewhere, to show the world it would not retreat. Grenada became the answer.

The Night Before

On the night of October 24, helicopters lifted off from carriers and airfields, loaded with men who only days earlier had been in garrison. Rangers packed rucks with extra ammunition. Marines checked weapons in dim red light aboard the ship. SEALs slipped into boats for predawn insertions.

Most of the young soldiers and Marines could barely find the island they would be fighting for on a map and had never seen combat. But by sunrise on October 25, Grenada would be etched into their memories. And once again, as in Vietnam, helicopters would decide who lived through the day.

Point Salinas: The Rangers' Ordeal

The unfinished runway at **Point Salinas** was an easy target on the maps spread across the briefing tables. It was open ground, flat and bare, with nowhere for an enemy to hide. But the hills above the airfield told a different story. Grenadian troops and Cuban advisers had dug in, and when the first helicopters came low on the morning of **October 25, 1983**, they were waiting.

Rangers of the **1st and 2nd Battalions** were packed into the new UH-60 Black Hawks. For many pilots, this was their first time flying the aircraft into combat. The plan called for a fast-rope assault directly onto the runway, but the plan unraveled before the first rope hit the ground.

Heavy machine guns and 23mm cannon fire erupted from the ridges. **Chief Warrant Officer Michael Durant** felt his aircraft buck as rounds smashed into the nose. Warning lights screamed in his cockpit. Durant forced the helicopter down on the edge of the strip, skidding into dirt and brush. His crew scrambled clear, pulling Rangers with them as the aircraft burned, crew chief **Sgt. Donald Jones** dragged one of the gunners by his web gear, shielding him with his own body until other Rangers got them both under cover. Durant survived, bloodied and shaken, but alive.

All along the line, helicopters took hits. **Capt. Keith Lucas** fought to keep his bird steady while his door gunner, **Sgt. Tom Brown**, fired until the barrel smoked. Brown later said, "It felt like we were inside a drum someone was beating with hammers.

The noise was constant. You just kept firing and hoped your guys hit the ground alive."

The Rangers who leapt out of the helicopters hit the runway hard. **Sgt. William Bass** remembered diving flat as tracers cut across the concrete. "There was no cover. You crawled forward on your belly and tried not to think about how far it was to the other side."

Others told of crawling past wounded men who urged them forward. A young Ranger from Charlie Company remembered a buddy gripping his sleeve, whispering, "Go, don't stop here." He pushed on, knowing medics would come when the helicopters returned.

Platoon leaders rallied their men. **Lt. John Abizaid**, later a four-star general, sprinted under fire to pull his Rangers into position. Men saw him waving them forward as rounds cracked around him, then dropping beside them to return fire. His courage steadied the line.

Close combat erupted at the far end of the runway. Grenadian and Cuban defenders fought from bunkers and trenches. Rangers tossed grenades, then rushed in firing M-16s and M-60s. **Cpl. James Flanagan** described it later: "You threw a frag, went in shooting, and hoped you were faster than the man in the hole."

As the fighting moved into the bunkers, others set up machine gun teams to cut the hillside. **Sgt. Paul Johnson** positioned his M-60 crew behind a dirt berm and laid down steady fire until the last defenders broke. His assistant gunner fed belts until his fingers blistered.

Behind them, the wounded were piling up. Rangers dragged them to the edges of the strip, setting them in rough rows for evacuation. Medics worked on the open ground, kneeling beside men while fire still echoed across the airfield. The first medevac helicopter dropped in low, its door gunner firing as the crew chief, **Spc. Robert Maynard** reached out and pulled in a bleeding Ranger by his harness. The Huey clawed away, trailing smoke.

By midday, Point Salinas was in American hands. The price was burned helicopters, smoking wrecks, and Rangers lying in blood on the concrete. Durant's downed Black Hawk smoldered on the edge of the field, a symbol of how costly the "light resistance" had been.

Years later, Rangers did not remember the briefings or the intelligence estimates. They remembered crawling across the runway with tracers snapping over their heads, the smell of smoke and fuel, and the sound of the Black Hawks circling back until the last squad was down. One Ranger put it: "I thought I was dead on that airfield. The only reason I'm here is because those pilots never quit on us."

Marines at Pearls and Seagle's Final Flight

While Rangers fought their way across the runway at Point Salinas, Marines of the **22nd Marine Amphibious Unit** approached the northern end of Grenada. Their mission was to secure **Pearls Airport** and nearby villages. For many of the

Marines, this was their first combat operation. For their pilots, veterans of Vietnam and younger men alike, it was another proof that helicopters would carry them into the teeth of fire.

The Marines flew in aboard **CH-46 Sea Knights** and **CH-53 Sea Stallions**. The twin-rotor Sea Knight was a familiar sight from Vietnam, a medium-lift helicopter used for decades to carry Marines into tight landing zones. The larger Sea Stallion had more power, more lift, and a reputation for muscle. Both would be tested that morning.

Seagle's Last Flight

Among the pilots was **Capt. Jeb Seagle**, at the controls of a CH-46 leading a group into Pearls. The airstrip stretched along Grenada's northeast coast, its approaches flanked by hills. As Seagle's helicopter came in low, gunfire erupted from entrenched positions. Anti-aircraft rounds smashed into the Sea Knight. Hydraulics failed, alarms screamed, and the aircraft dropped short of the strip.

Seagle stayed with the controls, fighting to keep the nose up and give his crew and passengers seconds to brace. The helicopter hit hard. Marines scrambled clear through smoke and fire. Witnesses later said Seagle's hands never left the stick until the bird was down. He was mortally wounded, but the seconds he bought saved lives.

Lance Cpl. Tom Gibbons, one of the riflemen who survived, put it: "He never quit. That bird was coming apart, but he held it level long enough for us to get out. I owe him my life."

Seagle was posthumously awarded the **Navy Cross**. For the Marines who walked away from his wreck, his name became a touchstone for courage.

The Fight at Pearls Airport

Other helicopters came in behind Seagle's. **Maj. Larry Taylor** brought his CH-53 into a landing zone already alive with fire. Bullets punched through the fuselage. A fuel line split, spraying mist into the cabin. His crew chief, **Sgt. Mike Herrera** wrapped it with rags while still pulling Marines aboard. Taylor kept the bird steady, lifted it heavily, and limped back to the sea. He turned around and went back in.

On the ground, Marines of **Company B, 1/8**, secured the airstrip and pushed into the surrounding area. Resistance was scattered but sharp, Cuban engineers fought from prepared positions, rifles in their hands instead of shovels. Marines cleared them out trench by trench.

One squad leader, **Sgt. David Long** remembered moving through scrub brush under fire. "One Marine from Bravo Company admitted later, "Most of us had never heard a shot fired in anger. We were green, no doubt about it. But when the helicopters came back again and again, dropping ammo and pulling our wounded out, it gave us backbone. If the pilots had that much guts, we had no excuse to freeze up."

The helicopters did not stop once Pearls were in American hands. They came in low with resupply, their rotors churning the dust, and they lifted out the casualties as fast as Marines could carry them aboard. Crew chiefs pulled men in by their web gear, their gloves soaked in blood. The cabins smelled of cordite, JP-4, and sweat. One remembered wiping down the deck plates with rags between runs to give the corpsman space to kneel by the next man who came in. Door gunners leaned into their weapons, laying down suppressive fire as Marines scrambled aboard.

The fight in the north did not last as long as the one at Point Salinas, but it carried the same weight for those who lived through it. Marines never forgot the sight of Seagle's Sea Knight burning at the edge of the strip, or the knowledge that he had held her level long enough to give them a chance.

Memory of a Marine

Years later, at reunions of the 22nd MAU, Marines spoke Seagle's name with respect. He had been one of them, and he had gone down doing what Marines expect of their leaders — staying with the aircraft, staying with the mission, and putting his men first.

Special Operations Baptism

While Rangers and Marines fought at the airfields, America's special operations forces faced their own trials on Grenada. For the **160th Special Operations Aviation Group**, later known as the "Night Stalkers," it was their first combat deployment. They flew **MH-60 Black Hawks** and **MH-6 Little Birds**, carrying SEALs and Delta Force operators to precision targets across the island. The missions were high-risk, the intelligence shaky, and the results often bloody.

Richmond Hill Prison

One of the most hazardous assignments was the attempt to insert operators at **Richmond Hill Prison**, a fortress situated on steep terrain. It was believed that political prisoners were being held there. To reach it, helicopters had to approach along a narrow corridor, climbing steeply toward walls lined with guns.

CWO Larry Bramlett piloted an MH-60 into the approach. The terrain left little room for maneuver. As his aircraft crested the ridge, anti-aircraft guns opened up. A burst tore through the fuselage, and the helicopter slammed into the hillside. Bramlett and his crew survived the crash, crawling out under fire and rallying with Delta troopers who fought to cover them. Bramlett would later say, "The prison was a fortress. We went in blind. It was a miracle more of us didn't die there."

Another pilot, **Maj. Larry Sloan** recalled pulling away from the target with smoke trailing from his engines. His crew chief fired an M-60 out the door until the belt ran out of ammunition. Sloan later received a Silver Star for bringing his helicopter and his men home.

Delta's Frustration

For Delta operators, Grenada was a bitter experience. They had been tasked with seizing critical sites, but poor maps and inadequate communications left them vulnerable. One Delta trooper described it as "fighting with one hand tied behind your back." At Richmond Hill, they fought their way clear without achieving their objective, carrying wounded with them as Night Stalker crews risked repeated runs to pull them out.

SEAL Team 6

The Navy SEALs paid dearly. On the night of October 24, before the main landings, a detachment from **SEAL Team 6** tried to infiltrate by sea to secure Grenada's radio station. The waters were rough, and their boats overturned in the surf. Four SEALs drowned before they ever reached the island: **Marc Anderson, Kenneth Butcher, Kevin Lundberg, and Stephen Morris.**

For their teammates, Urgent Fury began with a loss. One SEAL who survived said later, "We never even made it to the fight. We grieved for our friends lost at sea and then had to keep going. That's the part no one talks about."

Lessons Learned

Grenada showed the courage of special operators and aviators but also exposed flaws. The Night Stalkers proved their skill, flying into zones that seemed impossible, yet intelligence failures left them paying a high price. SEALs and Delta fought bravely, but the lack of coordination between services left them isolated.

Still, the men who were there remembered the bonds, not the politics. Bramlett remembered crew chiefs hauling wounded into smoking helicopters, refusing to leave men behind. A Delta operator remembered the pilots circling back into fire when everyone knew they had already been hit once. "They didn't ask if it was smart," he said. "They just asked where we were and came back."

The wounded stacked up fast on Grenada. Rangers at Point Salinas and Marines at Pearls repeatedly called for helicopters.

The name "Dustoff" was no longer painted on the side of the aircraft, but the spirit remained the same. Crews went into the fire to rescue the men.

Capt. John Adams, flying a UH-60, was told not to land on the runway because heavy guns still covered it. He landed anyway. "We could see the wounded lying out there. If we waited, they were gone," he said later. Rangers hauled casualties aboard while his crew chief, **Sgt. Richard Howell** pulled one man in by his web gear as blood soaked through his trousers. Howell used his own belt as a tourniquet and shouted over the noise for the medic to start administering fluids.

Other crews faced the same storm. **Chief Warrant Officer Rick Boyle** brought his helicopter low enough that rounds punched through both doors: his door gunner, **Spc. David Adams** leaned out and fired until the last belt of ammunition ran dry. The crew dragged three Rangers in, one shot through both legs, another hit in the chest. Boyle coaxed the helicopter out, heavy and limping.

Inside the cabins, medics worked on deck plates slick with blood. They braced with knees and elbows, holding IV bags in one hand and pressure dressings in the other. A pararescueman who rode in on one sortie remembered, "It was loud and violent, the bird shaking the whole time. I had to yell right in a man's ear to tell him he was going to make it."

Crew chiefs carried their own memories. One recalled wiping blood off the floor with rags after every flight, to give the corpsman space to kneel. Another said the smell of JP-4, mixed with cordite and sweat, lingered in his nose for months.

By the end of the week, medevac pilots had flown dozens of sorties, many of them under fire. Several aircraft returned with bullet holes patched by speed tape on the-metal panels. A few never came back at all.

The Aftermath

Operation Urgent Fury lasted just over a week. American forces captured the coup leaders, rescued the medical students, and took control of the Cuban-built airfield. The White House called it a clean victory, a sign that the United States had shaken off the doubts of Vietnam.

For the men who had fought there, the memories were less tidy. Nineteen Americans were killed, more than a hundred wounded, and helicopters were scarred or lost. At Arlington, families buried soldiers and Marines who had died on a mission that many Americans had only learned about when the operation was already over.

President Reagan told the nation, "Our armed forces showed once again that they are the best in the world." In Grenada, Rangers remembered crawling across the airfield under fire. Marines remembered pulling survivors from Capt Jeb Seagle's wreck. SEALs remembered the names of the four men who drowned before the invasion even began.

Lessons Written in Blood

Grenada exposed the seams in America's ability to fight in joint operations. Radios often failed. Army, Navy, and Air Force units could not talk to one another. Intelligence had underestimated Cuban and Grenadian defenses. Commanders argued about who was in charge.

Yet the helicopters never stopped. Pilots like Michael Durant and John Adams, crew chiefs like Richard Howell, and Marines who lived because of Jeb Seagle's final flight gave Urgent Fury its lasting meaning.

A Ranger summed it up years later: "We didn't win Grenada because the plan was perfect. We won because the pilots came back for us, no matter how hot it was. That's what got us through."

Chapter Notes –Operation Urgent Fury, Grenada 1983
Primary Sources:

After Action Report, Operation Urgent Fury, XVIII Airborne Corps, October–November 1983.

U.S. Marine Corps Command Chronology, 22nd Marine Amphibious Unit, October 1983.

U.S. Army Center of Military History, *Urgent Fury: Grenada 1983* (monograph).

Official Histories:

Ronald H. Cole, *Operation Urgent Fury: The Planning and Execution of Joint Operations in Grenada, 12 October–2 November 1983* (Joint History Office, Office of the Chairman of the Joint Chiefs of Staff, 1997).

U.S. Army Combat Studies Institute, *The Ranger Experience in Grenada, 1983*.

U.S. Naval History and Heritage Command, reports on naval aviation during Urgent Fury.

Published Accounts:

Mark Adkin, *Urgent Fury: The Battle for Grenada* (Lexington Books, 1989).

Hugh Shelton & Ronald Cole, *The Joint Lessons of Grenada* (Military Review, 1985).

Veteran Testimonies:

Interviews with Chief Warrant Officer Michael Durant (Black Hawk pilot at Point Salinas), Capt. Keith Lucas, and crew chief Sgt. Tom Brown, published in the *Army Aviation Digest*, 1984.

Oral histories of Rangers from 1/75 and 2/75 recorded by the U.S. Army Ranger Association.

Marine accounts of Capt. Jeb Seagle's last flight preserved in 22nd MAU reunion histories.

SEAL Team 6 tributes to Marc Anderson, Kenneth Butcher, Kevin Lundberg, and Stephen Morris, maintained in SEAL memorial records.

Clarification:

Operation Urgent Fury officially lasted one week, but planning and aftermath extended longer. While many Americans remember it as a "clean victory," survivors stressed the intensity of fire, the confusion of communications, and the sacrifice of helicopter crews and infantry on the ground.

CHAPTER EIGHTEEN

Operation Just Cause: Panama, 1989

In December 1989, Panama was a country on edge. **General Manuel Noriega** ran it with a mixture of fear, cocaine profits, and military muscle. Once he had been helpful to Washington, even drawing a salary from the CIA, but by the late 1980s, he had turned into a liability. His power was maintained by corruption and intimidation. His soldiers, the Panamanian Defense Forces, stopped cars at will, shook down civilians, and harassed American servicemen and their families who lived in the Canal Zone.

Inside Panama, some still called him *El Hombre* and treated him as untouchable. Others whispered that he had gone too far. The underground press published stories of disappearances. In neighborhoods of Panama City, people said the walls had ears, because Noriega's men were everywhere.

The Breaking Point

The situation boiled over on **December 16, 1989**. A car carrying four unarmed U.S. Marines approached a PDF roadblock in Panama City. The Marines were stopped and surrounded. In the confrontation, **Lt. Robert Paz**, just 25 years old, was shot and killed at point-blank range. Another Marine was severely beaten.

The news raced through U.S. bases. Marines who had served with Paz remembered him as quiet and steady. One buddy said, "Rob was the kind of officer you wanted to follow. To see him murdered like that, it lit a fire in all of us."

Families in the Canal Zone were rattled. Wives were told to keep their children indoors. Some recalled rocks thrown at their cars by Noriega's "Dignity Battalions," the armed civilian militias loyal to his regime. For many, it felt like the country was spiraling into lawlessness.

Washington Responds

President George H. W. Bush met with his advisors within hours of the incident. The protection of the Panama Canal, the safety of American citizens, and the removal of Noriega all came to the forefront. Bush's words were blunt: *"The lives of*

American citizens are in danger. We must protect the integrity of the Panama Canal and uphold democracy."

The Pentagon began planning what would become the largest U.S. combat operation since the Vietnam War. The mission would remove Noriega, safeguard the canal, secure American families, and install Panama's rightful government. The operation was given a name meant to reflect both legality and force: JUST CAUSE.

More than **27,000 U.S. troops** would be committed, backed by 300 aircraft. The **82nd Airborne Division** prepared for mass parachute drops into Panama City. The **75th Ranger Regiment** rehearsed airfield seizures. The **7th Infantry Division (Light)** staged for urban combat. Special operators — **SEALs, Delta, and Task Force 160 Night Stalkers** — received missions to hit Noriega's command posts, airports, and strongholds.

Helicopters would carry much of the burden. Black Hawks, Little Birds, and Chinooks would deliver assault troops to rooftops, airfields, and city streets. Medevac crews would follow close behind, ready to pull casualties from the fire.

In the weeks leading up to the invasion, tension escalated within Panama. U.S. citizens reported harassment daily. Soldiers and families in the Canal Zone were briefed to expect mob attacks. One Army spouse later said she kept her children home from school because she heard that crowds were planning to rush the gates of Fort Clayton.

Panamanians who opposed Noriega watched quietly, knowing the United States was preparing something. One shopkeeper in Panama City said, "We could feel the storm coming. Soldiers in the street, foreigners looking nervous, whispers everywhere."

The Night Before

On **December 19**, the storm broke. Soldiers and Rangers were told to load. Helicopter crews moved by flashlight across flight lines, running pre-checks while crew chiefs tapped panels and loaded ammunition belts into guns. Rangers strapped on parachutes and checked each other's gear with nervous jokes.

Most were young, many still teenagers. They had been told the PDF would fold quickly, that resistance would be scattered. Some believed it. Others weren't so sure.

As the night deepened, Black Hawks, Chinooks, and transports lifted into the dark sky. In hours, they would drop or land thousands of soldiers across Panama City. It would be the largest combat air assault since Vietnam, and for many of those young men, it would be the first time they ever heard shots fired in anger.

The Airborne and Ranger Assaults

In the first hours of **December 20, 1989**, the skies over Panama filled with aircraft. It was the most significant U.S. combat drop since World War II. Rangers of the **75th Regiment** and paratroopers of the **82nd Airborne Division** descended into darkness lit by tracers and burning fuel. Their target was **Torrijos-Tocumen International Airport**, the gateway into Panama City.

The drop was planned for speed. Thousands of parachutes blossomed against the faint dawn, men hitting the ground with weapons clattering around them. The first Rangers hit hard, rolled, and scrambled to their feet as Panamanian Defense Force troops opened fire.

Sgt. First Class Tom Healy, a Ranger squad leader, remembered the shock. "We expected a few rounds, not anti-aircraft guns shooting at us while we were still in the air. I hit the ground and thought I'd landed in the middle of a fireworks show."

As Rangers secured the runways, paratroopers from the **82nd Airborne** poured in behind them. **Pfc. James Hathaway** later told an interviewer, "When my chute opened, I could see tracers passing underneath. You just prayed you wouldn't hang in the harness too long."

Black Hawks and Little Birds

While the paratroopers dropped, helicopters carried assault teams into the fight. Black Hawks of the **160th Special Operations Aviation Regiment** flew low over the city, inserting troops onto rooftops and beside government buildings. **MH-6 Little Birds** carried soldiers on side benches, skimming treetops before flaring onto narrow streets.

One Black Hawk, flown by **Capt. Jack Keane** landed Delta operators near PDF headquarters. Fire erupted from windows as soon as the skids hit, crew chief **Sgt. Alan Ritchie** leaned out, firing bursts from his door gun as operators sprinted into the compound. Keane pulled away trailing smoke, returned to refuel, and went back for another load.

The Airport Fight

At Torrijos-Tocumen, the firefight raged across the runways. PDF soldiers fired from hangars and control towers. Rangers moved building to building, clearing with grenades and bursts from M-60 machine guns.

Lt. Col. Wesley Taylor, who had led Rangers at Point Salinas six years earlier, now commanded 2nd Battalion, 75th Rangers. He remembered seeing men scattered across the field, rallying into squads under fire. "It was chaos, but the training held. You could see sergeants taking charge, pulling groups together, and pushing forward."

By midmorning, the airport was secured. Black Hawks began shuttling in reinforcements, landing heavy with troops and lifting out the first waves of wounded.

Elsewhere in Panama City, combat turned block by block. Paratroopers of the **82nd Airborne** fought through the crowded neighborhoods of Rio Abajo and San Miguelito. Helicopters circled overhead, their rotors beating down dust as they inserted squads at intersections.

Spc. David Harper, nineteen years old, recalled, "It was my first firefight. The Black Hawks dropped us into the middle of a street, and as soon as we hit the ground, rounds were snapping off the walls. I'll never forget that sound."

The first day was costly. Helicopters came back riddled with holes. Door gunners carried the memory of rounds sparking off their armor plates and ricocheting through the cabin. Medics treated chest wounds and shattered legs on vibrating deck plates as pilots coaxed their birds away from fire.

Capt. James Stone, a flight medic, said later, "We worked in blood seemingly ankle-deep. We'd land, pull three men in, take off, patch what we could, and then go back. It never stopped until dark."

By the end of the first day, U.S. forces held the major airfields, but Panama City was still burning. The invasion was underway, and helicopters were proving once again that they were not just transport but lifelines.

Special Operations in Panama City

As Rangers and paratroopers secured the airfields, other units were moving into the city itself. The missions fell to SEALs, Delta operators, and the aviators of Task Force 160. They went after Noriega's strongholds, the command posts, intelligence centers, and the places where his loyalists were dug in. Most of these teams arrived by helicopter, some by boat, and they found themselves in running gunfights almost from the moment they landed. The opening hours of Just Cause were confused and violent, and every man who went in knew he was stepping straight into the heart of the regime's defenses.

The SEAL Missions

The Navy SEALs were tasked with some of the most dangerous jobs. **SEAL Team 4** and **SEAL Team 6** were ordered to secure **Paitilla Airfield**, where Noriega's private aircraft was parked. Intelligence warned that the Panamanian Defense Forces would defend it heavily.

The SEALs launched before dawn, moving in small boats and helicopters. As they approached, heavy fire erupted. One SEAL, **Petty Officer First Class Chris Tilghman**, remembered it as "a wall of lead." Boats were riddled with bullets. When the team hit the beach, they were already taking casualties.

Four SEALs, **Lt. Junior Grade John Connors, BM1 Chris Tilghman, HM1 Don McFaul, and BM2 Issac Rodriguez**, were killed in the assault, heavy losses in the small SEAL community. McFaul's sacrifice stood out. He had already reached cover but ran back into the open to pull a wounded teammate to safety. He was shot and killed in the act. His shipmates later said, "He tried to save lives with his last breath."

Despite the losses, the SEALs held the airfield and denied Noriega his escape route.

Delta Force Raids

Delta operators went in against Noriega's military headquarters and intelligence facilities. They rode in on **160th SOAR Black Hawks**, landing on rooftops and in tight urban clearings.

Capt. Gary O'Neal, one of the assault leaders, described the ride in: "We were skimming treetops, and you could see tracers climbing past the rotors. The pilots never flinched." When they landed, operators sprinted for the doors, blasting through with explosives. Inside, they fought room to room, pulling hard drives, files, and radios.

At one site, a Delta trooper was hit in the leg. His teammate dragged him back to the helicopter while the crew chief leaned out, firing his M-60 down the street. The Black Hawk lifted away, heavy, rounds slamming into its fuselage, but the man survived.

The Night Stalkers' Role

For the **Night Stalkers**, Panama was their second major test after Grenada. Their Black Hawks and Little Birds flew dozens of sorties on the first day alone.

One Little Bird pilot, **CW3 Bill Jacobs**, landed his MH-6 on a narrow street, rotors barely clearing the rooftops. Operators jumped from the side benches and kicked in the doors of a PDF safe house. Jacobs held the bird steady until the last man jumped, then pulled away under fire.

Another crew, flying an MH-60, took an RPG hit near the National Assembly building. The explosion ripped through the tail, but the pilot, **Maj. David Alvarez** kept the helicopter flying long enough to crash-land in the Canal Zone. All aboard survived, thanks to his control in those last moments.

Medevac Under Fire

As special operators pushed through the city, medevac helicopters followed. **Capt. Roger McDonough**, an Army UH-60 pilot, landed twice under direct fire to pull wounded SEALs from Paitilla, crew chief **Sgt. Mike DeAngelo** leaned out of the doorway, dragging a bloodied man aboard while tracers whipped past. McDonough said later, "We weren't leaving them on that strip. Not while we had fuel and rotors turning."

A Bloody Morning

By sunrise, Panama City was echoing with gunfire and explosions. SEALs mourned their dead, Delta carried off wounded, and Night Stalkers counted bullet holes in nearly every bird they flew. Yet the missions had succeeded. Noriega's airfields were denied, his headquarters raided, and American control of the skies established.

A SEAL summed it up years later: "We paid dearly at Paitilla. But when the helicopters kept coming, even into heavy fire, it told us that no matter how bad it was, we weren't alone.

Medevac and Aftermath

By the second day of **Operation Just Cause**, the fighting in Panama City had taken a heavy toll. Paratroopers, Rangers, SEALs, and Delta operators all carried wounded back to landing zones, where helicopters beat down the dust and smoke to haul them away.

Capt. Roger McDonough, a UH-60 pilot, flew repeated medevac missions under fire, one of his crew, **Sgt. Mike DeAngelo** remembered dragging a bloodied SEAL aboard at Paitilla while tracers snapped through the cabin. "We weren't leaving them there," DeAngelo said. "Not while we had blades turning."

Another flight medic recalled working on deck plates slick with blood. He braced himself against the cabin wall, holding an IV bag in one hand and keeping pressure on a chest wound with the other. "You learned quickly that you couldn't hear your own voice," he said later. "The only way to let a man know he was alive and was going to stay that way was to put your hand on his shoulder and keep working."

Casualties and Courage

The first week of the invasion cost the United States **23 killed and more than 300 wounded**. For helicopter crews, it was a blur of missions. Birds came back shot through, patched overnight, and sent back up at dawn. Crew chiefs counted holes, shrugged, and went back to loading ammo crates.

Spc. David Harper, who had been nineteen when he hit the streets on the first night, later said what kept him going was the sound of the helicopters circling above. "You heard that thump in the dark, and you knew somebody was still coming for you," he recalled.

The Fall of Noriega

Noriega fled his headquarters and went underground. American forces hunted him through Panama City while helicopters ferried troops into likely hiding places. Eventually, he took refuge in the Vatican embassy. For weeks, U.S. helicopters circled overhead blasting rock music through loudspeakers, psychological warfare to unnerve him. Noriega finally surrendered on **January 3, 1990**.

For the soldiers and Marines who had fought in the streets, his capture was anticlimactic. They remembered the firefights, the heat, the smell of fuel in the cabins of Black Hawks, and the friends they had loaded onto litters. The politics felt distant.

For Washington, Just Cause was declared a clean victory. The Canal was secure, Noriega was in custody, and American credibility was reaffirmed. For those who flew and fought, the memories were more complicated.

A Night Stalker pilot said years later, "We had better gear than Grenada, but the same problems. Bad maps, bad comms, and you solve it with guts."

For the medevac crews, the mission was a continuation of Kelly's creed in Vietnam. They went in because men were bleeding, not because it was safe. One Ranger put it best: "The plan wasn't perfect. But the helicopters never quit on us. That's the reason we're here to tell it."

Chapter Notes Operation Just Cause, Panama 1989

Primary Sources:

After Action Report, XVIII Airborne Corps, *Operation Just Cause*, December 1989 – January 1990.

U.S. Army Center of Military History, *Operation Just Cause: The Incursion into Panama, December 1989–January 1990*.

U.S. Southern Command briefing papers, December 1989.

Official Histories:

Ronald H. Cole, *Operation Just Cause: The Planning and Execution of Joint Operations in Panama, December 1989–January 1990* (Joint History Office, 1995).

U.S. Army Special Operations Command monographs on Ranger, Delta, and SEAL operations during Just Cause.

U.S. Army Combat Studies Institute, *Airborne Assault on Torrijos-Tocumen International Airport*.

Published Accounts:

Thomas Donnelly, Margaret Roth, and Caleb Baker, *Operation Just Cause: The Storming of Panama* (Lexington Books, 1991).

Mark Bowden, *Guests of the* Ayatollah (sections covering U. S. perceptions of credibility and intervention strategy in the late 1980s).

News coverage from *Stars and Stripes*, *The Washington Post*, and *New York Times* (December 1989–January 1990).

Veteran Testimonies:

Interviews with helicopter pilots from the 160th SOAR, including CW3 Bill Jacobs and Maj. David Alvarez, archived in Army   Aviation Digest, 1990.

Oral histories from Rangers of 2/75, including recollections of Sgt. First Class Tom Healy and Lt. Col. Wesley Taylor, published in Ranger Association newsletters.

SEAL community tributes to HM1 Don McFaul, who gave his life pulling a teammate to cover at Paitilla Airfield, preserved in SEAL memorial archives.

CHAPTER NINETEEN

Operation Desert Storm, 1991

Prelude & Context

The invasion shocked Washington and its allies. Iraq now sat astride a considerable portion of the world's oil reserves. President George H. W. Bush promised the aggression would not stand. The United Nations condemned the invasion, imposed sanctions, and By the summer of 1990, the Persian Gulf had become the center of world attention. On **August 2, 1990**, Saddam Hussein's Iraqi Army stormed across the border into Kuwait. Within hours, armored divisions and mechanized infantry had overrun the small nation. Kuwaiti resistance collapsed in two days. Oil fields burned, civilians fled, and the world watched the spectacle on television.demanded Iraq's withdrawal. Saddam Hussein ignored them all.

The Enemy

Saddam bragged openly to the world's press. He promised that any attempt to drive him out of Kuwait would lead to *"the Mother of All Battles."* Iraqi television broadcast parades of armored vehicles and soldiers chanting his name. He boasted that the desert would become "a graveyard" for American soldiers.

His confidence was not empty. Iraq had the **fourth-largest army in the world** at the time, with more than a million men under arms. They were battle-tested from eight years of brutal war with Iran, a conflict that had cost more than a million lives. Iraqi forces were equipped with thousands of tanks, many of them Soviet made T 72s, and dug into defensive belts across Kuwait.

The **Iraqi Air Force** operated nearly 700 combat aircraft, including **MiG-29 Fulcrums, MiG-25 Foxbats, and Su-24 Fencers**, all of which were supplied by the Soviet Union. Many of their pilots had real combat experience against Iranian F-14s and F-4s. Saddam promised his air force would contest the skies and bleed coalition forces.

The Coalition Builds

The U.S. launched **Operation Desert Shield**, deploying troops to Saudi Arabia to block further Iraqi advances. Within weeks, the desert filled with American units: the **101st Airborne Division (Air Assault), 24th Infantry Division, 1st Cavalry Division, 82nd Airborne Division**, and Marines of the 1st Marine Expeditionary Force. British, French, Saudi, and Egyptian units joined them. Eventually, more than **30 nations** contributed to the forces. Helicopters poured into the theater alongside the infantry.

Black Hawks, Apaches, and Chinooks lifted from staging areas in Saudi Arabia, joined by Marine CH-46s and CH-53s along the Gulf. **CW4 Mike Durant**, a Black Hawk pilot with the 101st, recalled the strange rhythm of life in the desert. "We lived in tents that filled with dust storms, flew drills day and night, and sweated through everything. When you shut the blades down after a night run, you feel like you are melting into the sand."

The Long Wait

From August through January, soldiers trained, rehearsed, and waited. An infantryman with the 24th Division said, "We drilled until the rifles felt welded to our hands. Everyone knew it was coming. Nobody knew when."

Mechanics worked through the nights under floodlights, fighting sand that clogged intakes and wore down engines. Pilots flew endless night-vision rehearsals, practicing deep strikes and fast rope insertions in blackout conditions.

Capt. **Larry Stubblefield**, an Apache commander, remembered sitting in his cockpit on the Saudi border staring north. "We knew Apaches would be the first to cross. We were eager, but you could feel the weight of it. It wasn't just another mission. It was history starting."

The Shock and Awe

On **January 17, 1991**, after months of sanctions and failed diplomacy, the storm broke at 2:38 a.m. Baghdad time, American cruise missiles, F-117 stealth fighters, and B-52 bombers roared into Iraq. For the first time, the world saw **"shock and awe."** Televised images showed Baghdad lit with anti-aircraft fire, explosions rocking the capital.

Iraq's vaunted air force barely left the ground. Coalition jets swept them aside in days, many fleeing to Iran rather than face American F-15s.

That same night, Apache helicopters fired the first shots of the war. Crews led by **CW3 David Sawyer** and **CW4 Barry McDaniel** destroyed Iraqi radar sites on the border, opening the corridor for the air campaign. "When the radars went down," Sawyer said later, "it was like a door opening. After that, the sky belonged to us."

The air war had begun, and helicopters were already carving their place in it.

The Ground War: Air Assaults and the 100-Hour Campaign

By late February 1991, weeks of air strikes had hammered the Iraqi military. Tanks burned in neat rows, bunkers lay shattered, and radar sites were rubble. Coalition leaders judged the time right to move. On **February 24**, the ground war began. It would last just one hundred hours, but in that brief span, helicopters would prove themselves vital once again.

Apaches in the Lead

The first shots of the ground offensive belonged to the **AH-64 Apache**. Before dawn, Apaches slipped across the border in pairs, flying low and masked against the sand. Their job was to take out Iraqi early warning radars and pave the way for armored thrusts.

CW3 Dave Sawyer, who had led the strike on the first night of the air war, flew again. He recalled the strange silence as the desert slipped beneath him. "It was dark, no lights, no horizon. Then the targets popped on the scope. We fired, and the whole sky lit up."

Apaches fired **Hellfire missiles** that streaked low across the desert into bunkers and radar vans. Within minutes, entire sites went silent. A cavalry trooper from the 24th Infantry Division said later, "The Apaches cut the eyes out of the enemy. After that, our tanks rolled free."

The Left Hook

Coalition strategy hinged on General **Norman Schwarzkopf's** "left hook," a massive sweep west and then north through the desert to strike the Iraqi Republican Guard from behind. To move such a force required helicopters.

Black Hawks ferried infantry into blocking positions. **CH-47 Chinooks** carried artillery batteries, lifting howitzers and crews miles in a single hop. In some sectors, Chinooks carried fuel bladders slung beneath them, keeping armored columns supplied in the trackless desert.

Capt. Jeff Stewart, a Chinook pilot, remembered flying loads of artillery at night. "You flew blackout, goggles on, eyes burning from the sand. You set down, dropped the guns, and were back in the air before the dust even cleared."

Medevac in the Desert

Casualty evacuation was constant. Black Hawk medevac crews, often called "Dustoff" again by the men who depended on them, flew into minefields and firefights to pull out the wounded.

Sgt. Mark Long, a flight medic, described one mission near the Kuwait border. "We landed under fire. Two Marines were hit by shrapnel. We pulled them in, and I worked on them the whole way back. I'll never forget looking down and seeing the deck plates red. But they lived. That's why we were there."

Crews faced dust storms that turned the world into a brown wall. Pilots sometimes hovered blind, trusting their crew chiefs to guide them in. One copilot said, "I couldn't see the nose of the bird. We were flying on instinct and trust."

As the coalition swept north, the most brutal fight came against the **Republican Guard divisions**, Saddam's elite. Tanks dug in hull-down positions, artillery hidden in wadis, and soldiers battle-tested from the Iran-Iraq War.

Apache pilots went after them with Hellfires and rockets. One company of Apaches from the 101st destroyed more than thirty tanks in a single night. A pilot said later, "You'd line up on a target, fire, and it was just fireballs across the desert. But they shot back. You could see tracers climbing right past the cockpit."

The Hundred Hours

By **February 28**, it was over. The Republican Guard was shattered, Kuwait was free, and Saddam's army was streaming north in defeat. The road from Kuwait City to Basra became known as the **Highway of Death**, littered with burned-out vehicles struck by air and helicopter attacks.

For the soldiers who fought the ground war, the brevity was almost shocking. One infantry sergeant said, "We trained for months, sweated in the desert, and when it came, it was four days of chaos and then it was done."

For helicopter crews, the memory was different. They remembered endless nights of sorties, lifting troops, flying fuel, firing Hellfires, and pulling wounded out of fire. A Black Hawk pilot put it best: "The war was short, but for us it never stopped moving. If you weren't flying, you were fixing. And then you flew again."

Aftermath and Lessons

The war ended quickly, but for the men who flew and fought it, the memories did not. On the morning of **February 28, 1991**, word went down the line that the cease-fire had been called. Crews that had been flying almost nonstop were told to stand down. One Black Hawk pilot said later, "We still had birds on the line with bullet holes in them, rotors patched with tape, and they told us it was over. It didn't feel over. It just felt unfinished."

What Helicopters Did

For the first time, helicopters had become an integral part of every layer of the fight. **Apaches** stalked Iraqi armor, often at night, firing Hellfires that turned bunkers and tanks into fireballs. **Black Hawks** carried troops into blocking positions, ferried ammunition, and hauled the wounded out of minefields and trenches. **Chinooks** lifted artillery and heavy loads, sometimes landing in the middle of a sandstorm to drop their cargo. Marine **Sea Knights** and **Sea Stallions** put infantry ashore from the Gulf and kept them supplied when trucks could not cross the desert fast enough.

Crews remembered the environment as much as the enemy. Dust storms that grounded jets did not stop helicopters, though they often flew blind. Sand clogged engines and chewed up blades. Navigation at night across the flat desert was like flying over a sheet of paper, with no horizon and no features to trust.

CW4 Mike Durant, flying Black Hawks with the 101st, said, "Every mission started with the desert trying to kill you. Before a round ever came your way, you were already fighting the sand."

Losses and Mistakes

The ground war lasted four days, but it still resulted in losses. Several Apaches were shot down by anti-aircraft fire when they pressed deep. Some helicopters crashed in brownout conditions, pilots disoriented by the blinding sand.

The worst moment came on **February 27**, when two U.S. Air Force F-15s mistakenly shot down two Army Black Hawks carrying troops from the 101st Airborne. Twenty-six Americans were killed. A soldier from the unit said later, "We had survived the Iraqi fire, the sand, everything. Then our own people killed us. That's the part you don't forget."

Medevac Memories

Dustoff crews carried the human weight. **Sgt. Mark Long**, a flight medic, remembered lifting two Marines hit by shrapnel near Kuwait City. "We were on the ground for less than a minute. We hauled them in, and I kept pressure on chest wounds the whole way back. I was yelling in their ears, telling them to hang on. That's all you could do. Talk to them and work with your hands."

Crew chiefs guided pilots into landings they could not see, half out the door, waving hand signals while brown dust swallowed the aircraft. One said, "I couldn't see the nose of the bird, but the guys on the ground were waiting. You get them out or you die trying."

The Highway of Death

When the end came, it was brutal. As the Iraqis retreated north out of Kuwait, the main highway toward Basra filled with armor, trucks, and cars packed with stolen goods. Coalition aircraft caught them in the open. Apaches worked the edges of the column, rockets and Hellfires slamming into vehicles until the road was choked with fire and smoke. Later, some pilots said it felt different from the other missions, less like a fight, more like a beating. One Apache gunner remembered, "They weren't even shooting back. They were just trying to run." They were just trying to get home, and we destroyed them."

What Was Learned

Desert Storm was hailed as a clean victory, but the lessons were complex and nuanced. Some were stories from the past. Communications between services often failed. Maps were outdated. Friendly fire was a constant fear. Yet helicopters proved their worth again, moving, striking, and saving lives.

A Night Stalker pilot summed it up years later: "The Apaches showed what we could do to an armored army. The Black Hawks showed what we could do for our own people. We learned a lot, some of it the hard way. Everything in Iraq and Afghanistan later came out of those four days."

For veterans, the memory was not of technology or strategy. It was of the endless flying, the weight of dust, and the faces of the wounded on litters in the cabins. An infantry sergeant said, "For us on the ground, the war was four days. For the helicopter guys, it never stopped. If they weren't hauling us, they were hauling ammo. If they weren't pulling us out, they were fixing their birds in the dark. That's what I remember."

Saddam had promised the Mother of All Battles. What he got was a hundred hours of helicopters, tanks, and infantry smashing through his lines. For the men who flew it, Desert Storm was not clean or simple. It was a fight against sand, against fire, and against time, and it left lessons that shaped every war that came after.

Chapter Notes –Operation Desert Storm, 1991
· **Primary Sources:**

o   U.S. Army Center of Military History, *Certain Victory: The U.S. Army in the Gulf War* (1993).

o   U.S. Air Force Gulf War Air Power Survey, Volumes I–V (1993).

o Department of Defense After Action Reports, *Operation Desert Storm*, February–March 1991.

· **Official Histories:**

o Stephen A. Bourque, *Jayhawk! The VII Corps in the Persian Gulf War* (CMH, 2002).

o Richard P. Hallion, *Storm Over Iraq: Air Power and the Gulf War* (Smithsonian, 1992).

o U.S. Army Aviation Digest, Gulf War issues (1991–1992).

· **Published Accounts:**

o Tom Clancy, *Into the Storm* (1997) with Gen. Fred Franks.

o Rick Atkinson, *Crusade: The Untold Story of the Persian Gulf War* (1993).

o Robert Scales, *Certain Victory* (1993).

· **Veteran Testimonies:**

o Oral histories from Apache and Black Hawk crews of the 101st Airborne Division, collected by the Army Aviation Museum, Fort Novosel.

o Interviews with CW4 Mike Durant and other pilots, Army Aviation Digest (1991).

o Testimonies of flight medics like Sgt. Mark Long, preserved in Dustoff Association archives.

o Ranger and infantry recollections of helicopter support, U.S. Army Ranger Association newsletters.

· **Clarification:**

o The air war began on January 17, 1991, followed by the ground campaign on February 24.

o The ground war lasted 100 hours, ending February 28 with a coalition ceasefire.

o U.S. losses totaled 148 killed in action, with 458 wounded. Helicopters played a role in nearly every phase, from the first Apache radar strikes to medevac and resupply in the final hours.

CHAPTER TWENTY

The Battle of Mogadishu, 1993

Prelude & Context

In the early 1990s, Somalia was a failed state. It had been torn apart by famine and was ruled by warlords after years of civil war brought an end to the government's control. Armed factions fought for control of Mogadishu while ordinary people starved. By 1992, images of emaciated children filled Western television screens. The United States and the United Nations sent food, medical supplies, and troops to ensure continued relief efforts.

What began as a humanitarian mission quickly turned into combat. Convoys were ambushed, aid shipments were stolen, and peacekeepers were killed. The most powerful warlord in Mogadishu, **Mohamed Farrah Aidid**, saw the foreign presence as a threat to his hold on power. His militia fighters, armed with AK-47s and rocket-propelled grenades, roamed the city's streets and alleyways.

By 1993, the United States shifted from feeding Somalia to trying to break Aidid's grip. Special operations forces were sent in under the banner of **Task Force Ranger**. The force included Delta operators, Army Rangers, and the helicopter crews of the **160th Special Operations Aviation Regiment, the Night Stalkers.**

Task Force Ranger Arrives

Task Force Ranger arrived in Mogadishu in late August 1993. Their mission was simple on paper: capture Aidid's top lieutenants and weaken his network. In practice, it meant flying into one of the most hostile cities on earth.

The Rangers provided security on the ground. Delta Force operators carried out the direct snatches. The Night Stalkers flew them in and out. The aviators brought MH-60 Black Hawks for lift and fire support, as well as MH-6 Little Birds for close insertion into tight city streets.

The men who crewed those helicopters were veterans of Grenada, Panama, and Desert Storm. They knew the risk of flying into a city where every alley could hide an RPG team. **CW4 Michael Durant**, one of the Black Hawk pilots, said later, "We knew the city was hot. Every mission in Mogadishu meant going into a hornet's nest."

The Threat in the City

Aidid's militia had grown used to foreign aircraft overhead. They learned to target them with rocket-propelled grenades.

Convoys had already been hit. In June 1993, 24 Pakistani peace-keepers were killed in an ambush, proof that Aidid's men could strike hard.

American soldiers in Mogadishu lived in a fortified compound near the airport. They flew missions almost daily. Rangers remembered running to helicopters at all hours, sweat soaking through their uniforms in the sweltering heat of Somalia. Crew chiefs remembered long hours spent checking blades, patching bullet holes, and replacing hydraulics that had been shot out on the last run.

A Ranger from B Company later said, "The city felt alive. You could feel the eyes on you the minute you lifted off. Everyone down there had a weapon, and half of them wanted to use it."

Setting the Stage

Through September, Task Force Ranger ran a series of raids, capturing dozens of Aidid's lieutenants. The missions were short and violent. Helicopters roared in, Delta and Rangers hit their targets, and then everyone lifted out under fire. Each raid brought more hostility. Somali crowds hurled rocks, and women and children pointed at helicopters to guide gunners on rooftops.

On October 2, commanders received intelligence that two of Aidid's senior men would be meeting in a house in the heart of Mogadishu the next day. The order was given, Task Force Ranger would go in. For the men who strapped into Black Hawks and Little Birds on **October 3, 1993**, it was supposed to be another quick raid, in and out in under an hour.

Instead, it would become one of the bloodiest urban battles since Vietnam, a day when helicopters, Rangers, and Delta operators would be tested beyond anything they had faced before.

The Assault, October 3, 1993

On the afternoon of **October 3, 1993**, the order went down: Task Force Ranger would hit a house near the Bakara Market where two of Aidid's senior men were believed to be meeting. The plan was straightforward. Black Hawks would insert Delta operators fast-roping onto the target building. Rangers would secure the perimeter at intersections. Little Birds would drop assault teams into the streets nearby. The whole mission was expected to last under an hour.

The men were loaded onto helicopters at the airfield. It was hot, the air heavy with dust and fuel fumes. Rangers wore body armor, web gear, and carried M16s or carbines. Delta operators sat calmly, checking weapons and radios. Crew chiefs did final walk-arounds, patting panels, tugging straps, and slapping the fuselage with a gloved hand before climbing aboard.

The Launch

Engines roared to life. The Black Hawks lifted in formation, blades chopping the air into thunder. MH-6 Little Birds skimmed the ground, nimble and fast, carrying operators on the benches strapped to their sides. The city stretched ahead, a sprawl of low buildings, narrow streets, and markets crowded with people.

Super 61, flown by **CW3 Cliff Wolcott** and **CW3 Donovan Briley**, led the Black Hawks. **Super 64**, flown by **CW4**

Michael Durant, was stacked above and behind. Other Black Hawks, call signs Super 62 and 63, orbited to provide fire support with their door gunners.

Rangers remembered gripping their ropes as the helicopters came over the objective. "You felt the bird pitch and you knew it was time," one said. "Heart hammering, hands slick on the rope, and then you slid."

Fast-Roping In

At the target building, Black Hawks flared over the roof. Delta operators dropped first, sliding down ropes onto the flat roof. Rangers followed at intersections, their boots slamming onto the pavement as the crowds scattered.

One Ranger, **Pfc. Todd Blackburn** lost his grip on the rope from Super 67 and fell nearly seventy feet to the street. His leg was shattered, and he was barely conscious. Medics rushed to him as bullets cracked down the street. It was the first sign the mission was not going to be quick or clean.

Crowds that had run at first began to gather, then militia fighters opened fire from alleys and windows. Rangers sprinted to secure corners, taking fire from rooftops. Little Birds darted down the avenues, their miniguns hosing bursts into firing positions.

The Fire Builds

In the Black Hawks above, crew chiefs leaned out with M60s and miniguns, spraying at muzzle flashes in the buildings. **Sgt. Paul Howe,** a Delta team leader on the ground, said, "The city lit up. It felt like every alley had an RPG team waiting."

Super 62, flown by **CW3 Stan Wood** and **CW4 Gary Gordon**, provided cover. Door gunners fired until their ammo belts rattled empty. Helicopters circled tight, banking hard to stay above the target.

Durant, in Super 64, orbited above the action. "You could see the city boiling," he said later. "Every street had people, some running, some shooting. It was clear this wasn't going to be quick."

First Casualties

Within minutes, the plan unraveled. Rangers moving to block intersections came under heavy fire. One squad leader said, "You looked one way and saw kids throwing rocks, looked the other and saw muzzle flashes. You didn't know who was who."

Blackburn's injury forced a shift in the plan. A convoy was ordered to take him back to the base. The convoy rolled into narrow streets choked with traffic, drawing fire from every direction. Radios crackled with confusion.

Above, Little Bird pilots banked through sheets of tracers. One pilot said, "It was like flying down a hallway with people shooting out of every doorway. You just kept moving."

The Trap Springs

The militia had been waiting. Fighters poured into the streets, firing AKs and RPGs. Crowds swarmed intersections. Rangers reported being cut off in small groups. Door gunners leaned on their triggers to keep the enemy back.

Delta operators cleared the target building, capturing Aidid's men, but by then the battle was spreading across the city. What was meant to be an hour-long raid had turned into a running fight.

Durant circled in Super 64, watching as the streets filled. "You knew then," he said later, "this wasn't going to be another quick snatch. We were in it for real."

The Sky Over Mogadishu

From the air, the city looked like it was on fire. Black Hawks weaved to avoid RPGs streaking upward. Little Birds fired rockets down alleys. Rangers hugged corners and returned fire from doorways. The mission had only just begun, and already the casualty count was rising.

A Ranger private summed up the moment years later: "We went in thinking it was going to be fast. It didn't take long to realize we were in the middle of a battle that wasn't ending anytime soon."

The Downings

The raid unraveled when the first helicopter crashed to the ground. In Mogadishu's streets, helicopters were the lifeline, circling above, pouring fire into alleys, carrying men in and out. When they were knocked down, the fight changed completely.

Super 61

At 3:45 in the afternoon, **Super 61**, piloted by **CW3 Cliff "Elvis" Wolcott** and **CW3 Donovan Briley**, took an RPG hit in the tail. The Black Hawk bucked hard, lost control, and spun into the streets below. The impact shattered the fuselage. Wolcott and Briley died instantly.

Crew chiefs **Staff Sgt. Bill Cleveland** and **Sgt. Ray Frank** was alive but trapped. Rangers sprinted toward the wreckage, cutting through alleys under fire to reach them. One Ranger remembered the horror of finding the helicopter in pieces, the street packed with smoke, civilians swarming, and militia closing in. "It felt like the whole city was coming for that bird," he said.

Overhead, other Black Hawks orbited, their gunners firing to keep the crowds back. "We saw Elvis's ship go in. We knew it was bad," one pilot recalled. "We were already asking for the Quick Reaction Force to roll before the dust even settled."

Super 64

Minutes later, the nightmare doubled. **Super 64**, flown by **CW4 Michael Durant**, was hit by an RPG. The tail rotor tore away. Durant fought the controls, trying to level the bird. It spun and slammed into a street five blocks from Super 61.

The crash left the crew broken. Durant's back was fractured, and his leg was shattered. His crew, Spc. **Tommy Field, Staff Sgt. Bill Cleveland, and Sgt. Ray Frank** was alive for moments, but the militia closed fast. Small arms fire poured in. Field and Frank were killed in the wreck. Cleveland, gravely wounded, was dragged from the helicopter by the mob.

Durant was dazed, in pain, still firing a pistol when he could. "We hit hard, and I knew I was hurt bad," he remembered. "But the fight was right there at the windows. They were already on us."

Gordon and Shughart

Circling above in **Super 62**, Delta snipers **Master Sgt. Gary Gordon** and **Sgt. 1st Class Randy Shughart** saw the crash. They understood Durant and his crew were about to be overrun. Twice, they asked to be inserted to protect the site. Twice, they were told no. Finally, permission came.

Super 62 flared into a nearby street. Gordon and Shughart jumped out with only their rifles, sidearms, and a handful of magazines. They fought their way to Durant's Black Hawk, dragged him from the wreck, and set him behind cover.

Durant never forgot their composure. "They moved like men who had done this a hundred times. Calm, deliberate. They set security, checked me, and went back to fighting. It was like they had all the time in the world, even though the world was ending around us."

The two men held off wave after wave. They fired from broken walls, shifted positions, and reloaded from the downed crew's weapons when their own ammo ran low. Militia fighters closed in from every side, pouring AK fire into the street.

Durant remembered Gordon being hit first. Shughart pulled him into cover and kept firing. Moments later, Shughart was shot as well. Both were killed.

When the militia finally overran the site, Durant was the only American left alive. Captured and beaten, he was taken prisoner. He would spend eleven days in captivity before his release.

Recognition

The courage of Gordon and Shughart became legend. For giving their lives to save their fellow soldiers, they were posthumously awarded the **Medal of Honor, a reward that seems small for their kind of courage.** Their actions were seared into the memory of every man who fought in Mogadishu that day.

Durant said later, "They didn't have to go in. They asked to. They knew the odds. They came anyway. I'm alive because of them. That's the only reason."

The Streets Close In

With two helicopters down, the battle became a race to secure the crash sites. Rangers and Delta fought to reach Super 61 and hold it. At Super 64, only Durant remained, his crew gone, his protectors killed. Overhead, Night Stalker pilots circled through sheets of tracers, firing into alleys to keep the militia back, calling frantically for ground convoys to push through.

A Ranger later summed it up: "The minute those birds went down, the mission changed. It wasn't about the raid anymore. It was about survival, and it was about not leaving anyone behind."

The Night Battle and Rescue Convoy

The sun dropped low over Mogadishu, but the battle only grew fiercer. Task Force Ranger was scattered across the city, pinned down at intersections, and surrounded at the two crash sites. The Quick Reaction Force convoy, meant to link them together and bring everyone home, was already in trouble. The convoy rolled out from the airfield, comprising dozens of vehicles, including Humvees and trucks, packed with Rangers, Delta operators, and support personnel. The streets of Mogadishu were narrow, crowded, and twisting. Maps were poor, radios jammed.

Almost as soon as they entered the city, they came under fire. RPGs streaked into the column. Trucks were hit, tires shredded, engines smoking. Rangers fought from the backs of vehicles, firing into alleys as the convoy slowed and ground to a halt.

Sgt. Keni Thomas, riding in one of the lead Humvees, remembered, "We were supposed to be the cavalry. Instead, it felt like we had driven straight into a hornet's nest."

The convoy became lost in the city's maze. Drivers took wrong turns. Vehicles backed into alleys too narrow to pass. All the while, Aidid's militia pressed in, firing from rooftops and crowds.

Pinned Down

At the crash sites, Rangers and Delta fought from rubble and broken walls. At **Super 61's site**, they built a perimeter around the wreck, trading fire across the street. Casualties mounted. Medics worked in the dirt, kneeling over the wounded as rounds snapped overhead.

Staff Sgt. Matt Eversmann, a squad leader, later recalled, "You didn't think about anything beyond the corner you were holding. You just kept firing and hoped the helicopters stayed up there above you."

Above them, Black Hawks circled low, dropping ammunition and water in canvas bags. Door gunners leaned out, firing into alleys where muzzle flashes lit up. The pilots flew through curtains of tracer fire, banking hard and coming back again and again.

At **Super 64's site**, only Durant remained, now captured and dragged away. The militia swarmed the area, and the perimeter collapsed. Other Rangers and Delta were too far to reach him. They fought to hold their own positions, still under heavy fire.

The Night Battle

As night fell, the city lit with gunfire. Tracers streaked in every direction. Helicopters flew with their miniguns glowing red, firing down at fighters who poured into the streets.

Little Birds skimmed the rooftops, rockets slamming into alleyways to break ambushes. One pilot said later, "You couldn't tell the good from the bad half the time. All you knew was who was shooting at your guys."

On the ground, Rangers held out in small pockets. Some fought from courtyards, others from abandoned buildings. Ammunition ran low. At one point, a Black Hawk dropped cases of ammo directly into a street where a Ranger squad was pinned. They fired until barrels glowed.

The Quick Reaction convoy never made it through before nightfall. Vehicles were disabled and wounded filled the trucks.

Command finally pulled them back to the airfield, battered and bloodied. The men at the crash sites and strongpoints were left to hold out.

Helicopters became the only link. They flew constant circuits, dropping supplies and evacuating wounded when they could land. Crew chiefs hauled bloodied men aboard and shouted for lift. Pilots often flew back with holes stitched across the fuselage.

One flight medic remembered, "You looked down and saw the tracers climbing like ropes into the sky. You pulled wounded in with your hands and tried not to think about how many times that bird had been hit already."

Through the Night

The battle dragged through the dark. Rangers counted magazines, checked each other's wounds, and braced for the next assault. They had been fighting for hours without food or water. Some went without helmets, lost in the chaos of the first contact.

Sgt. John Stebbins later said, "You were down to your last rounds, and you still had to cover your buddy. It was just survival."

Above, pilots fought exhaustion. They had been flying since the afternoon, weaving through anti-aircraft fire, dodging RPGs, and watching friends go down. Yet they kept coming back, because if they did not, the men on the ground would be overrun.

At dawn, relief finally came. Pakistani tanks and Malaysian armored personnel carriers rolled through the streets, leading another rescue convoy. They smashed through barricades, their heavy armor shrugging off the fire that had crippled Humvees the night before.

The battered Rangers and Delta survivors climbed aboard, bloodied, exhausted, and nearly out of ammunition. Some had been fighting nonstop for 18 hours.

One Ranger described the moment: "We climbed into the armor, and it was the first time you felt like maybe you'd live. The helicopters had kept us alive through the night, but the armor finally got us out."

Aftermath and Lessons

By the time the battle ended on the morning of **October 4, 1993**, Task Force Ranger had been fighting for almost 18 hours straight. The price was heavy. Eighteen Americans were dead. More than seventy were wounded. Two Black Hawks had been shot down, and several more were damaged. Somali casualties were staggering, with hundreds of fighters and civilians killed in the streets.

Durant's Captivity

Michael Durant, pilot of Super 64, was the only American captured. Beaten and paraded before cameras, he became a bargaining chip in Aidid's hands. For eleven days, he was held in safe houses around Mogadishu, injured and guarded by militia fighters.

Durant later recalled the pain and the fear, but also the determination to survive. "I thought about my family. I thought about the men who had died trying to protect me. I made up my mind I would live for them." He was released on October 14 after negotiations, a thin and battered man who walked with help when he stepped back onto U.S. soil.

Medal of Honor

The courage of **Gary Gordon** and **Randy Shughart** was recognized with the **Medal of Honor**, the first such award since the Vietnam War. Their families accepted the medals from President Bill Clinton in 1994. Clinton told them, "Your loved ones went into harm's way knowing the odds, and they went anyway. America will never forget them."

For soldiers who knew the two men, the awards were more than a ceremony. A Delta operator who had fought at the crash site said, "They asked to go in when everyone else knew it was suicide. They went because we were down there, and that's all that mattered."

Impact on Aviation

For the helicopter crews of the 160th SOAR, Mogadishu was a turning point. It showed both their skill and their vulnerability. Black Hawks and Little Birds had flown through a storm of fire, delivering men, dropping supplies, and pulling out wounded. However, it also demonstrated how vulnerable they could be against RPGs in tight urban terrain.

After the battle, the Army poured effort into upgrading aircraft protection, night vision equipment, and communications.

Tactics for urban combat were rewritten with Mogadishu in mind.

One pilot summed it up: "We proved we could do the impossible. We also learned we needed better tools. Both things were true that night."

Policy Consequences

The battle also had significant political consequences. Images of American soldiers dragged through the streets shocked the nation. Public opinion turned sharply against U.S. involvement in Somalia. Within months, President Clinton ordered a withdrawal. For years afterward, the phrase "Black Hawk Down" became a shorthand reference to the risks of getting drawn into foreign conflicts.

Some soldiers felt abandoned by the decision. A Ranger officer said, "We lost men, and then we pulled out. It felt like we gave their lives away." Others believed it was the only choice. Either way, Mogadishu changed how leaders thought about deploying American forces.

For those who fought, the battle was not a political debate; it was a matter of life and death. It was the memory of the men beside them. Rangers remembered dragging wounded through alleys. Delta operators remembered fighting house to house with no rest. Helicopter crews remembered patching bullet holes, rearming, and going back into the sky again.

A flight medic put it plainly: "We didn't win or lose in Mogadishu. We just did our jobs, minute by minute, trying to keep people alive."

The names of the dead are etched in stone now, but for those who survived, the sound of Mogadishu still comes back in dreams, the crack of AK fire, the shriek of RPGs, and above it all the steady beat of rotor blades carrying men into a fight that none of them would ever forget.

Chapter Notes The Battle of Mogadishu, 1993

Primary Sources:

U.S. Army Center of Military History, *The United States Army in Somalia, 1992–1994.*

Department of Defense After Action Reviews, Task Force Ranger, October 1993.

160th SOAR flight logs and unit histories from Mogadishu operations.

Official Histories:

Joint Chiefs of Staff, *Joint Report on Operations in Somalia* (1994).

U.S. Army Special Operations Command Monograph Series: *Urban Operations in Mogadishu.*

Congressional Research Service, *Somalia: U.S. Involvement, 1992–1994.*

Published Accounts:

Mark Bowden, *Black Hawk Down: A Story of Modern War* (Atlantic Monthly Press, 1999).

Michael Durant and Steven Hartov, *In the Company of Heroes* (2003).

John Stebbins, *The Road to Mogadishu* (2004).

Veteran Testimonies:

Oral histories of Rangers, Delta Force operators, and Night Stalker pilots recorded by the U.S. Army Ranger Association and 160th SOAR veteran networks.

Interviews with CW4 Michael Durant describing the shootdown of Super 64 and his captivity.

Ranger testimonies collected at reunions, including Staff Sgt. Matt Eversmann, Sgt. Keni Thomas, and others from B Company, 3/75 Rangers.

Delta operators' accounts of Gordon and Shughart's actions, recorded in SOF community memorial archives.

Clarification:

Task Force Ranger lost 18 men killed and more than 70 wounded. Somali militia casualties are estimated in the hundreds.

Gordon and Shughart were posthumously awarded the Medal of Honor, the first since Vietnam.

Images of U.S. soldiers killed and dragged through Mogadishu streets had a major impact on U.S. foreign policy, leading to withdrawal from Somalia in 1994

Chapter Twenty-One

Afghanistan, 2001

On the morning of **September 11, 2001**, Americans watched in horror as hijacked planes struck the World Trade Center and the Pentagon. Nearly 3,000 people were killed. Within hours, the United States knew who was responsible: **Osama bin Laden** and the al-Qaeda network sheltered by Afghanistan's Taliban regime.

President George W. Bush declared, "We will make no distinction between the terrorists who committed these acts and those who harbor them." The Pentagon began planning a campaign unlike any it had waged before.

Afghanistan was a remote, landlocked country ruled by a militia hardened by decades of war. Soviet forces had once lost tens of thousands of men trying to hold the same ground. Now American commanders would send small teams of Special Forces and CIA paramilitaries into mountains where roads barely existed. Helicopters would be the only means of access in and out.

Task Force Dagger

The first units tasked to go in were part of **Task Force Dagger**, comprising Army Special Forces and CIA teams. They would link up with Northern Alliance fighters resisting the Taliban. To insert them meant flying deep over high mountains at night, often hundreds of miles from friendly bases. The mission fell to the **160th Special Operations Aviation Regiment, the Night Stalkers**.

Night Stalker crews had already earned their reputation in Grenada, Panama, and Somalia. Afghanistan would test them like never before. They would have to fly **MH-47 Chinooks** and **MH-60 Black Hawks** across the Hindu Kush, a mountain range where peaks reached heights of more than 20,000 feet, and winter winds tore through valleys like knives.

CW3 Greg Calvert, a Chinook pilot, remembered the first rehearsals. "We knew the altitudes alone would push the aircraft to the edge. You add the weather, the distance, and the fact that the enemy was dug in, and you had everything stacked against you."

America at War Again

By October, carriers in the Arabian Sea were launching bombers into Afghanistan. Cruise missiles struck Taliban camps. However, airstrikes alone would not be enough to topple the regime. U.S. forces had to be on the ground.

On **October 19, 2001**, Rangers of the **3rd Battalion, 75th Ranger Regiment** made the first large-scale assault of the war. Carried by Chinooks, they struck **Objective Rhino**, a Taliban airfield south of Kandahar. The helicopters came in low under the cover of night, dust clouds rising as ramps dropped and Rangers charged out.

Sgt. First Class Matt Eversmann, who had fought in Mogadishu eight years earlier, watched from a command post as the reports came in. "The Night Stalkers dropped those boys right into the lion's den. Dust everywhere, gunfire, chaos. And they held the strip by dawn."

Voices From the First Days

Marines moved in soon after, establishing **Camp Rhino** as a forward base. CH-53 Sea Stallions carried them in heavy loads. One Marine corporal said, "We knew this wasn't going to be a short fight. The minute those helicopters lifted off, we were in it for the long haul."

CIA officer **Gary Schroen**, leading one of the first teams into the Panjshir Valley, later recalled looking out the back of a Chinook as it threaded through the mountains. "It was black, no lights, just the sound of the rotors and the cold air cutting in. You knew once you landed, it was you, your Afghan allies, and whatever you carried on your back."

The War Opens

Within weeks, U.S. forces and their Northern Alliance partners had pushed the Taliban out of Kabul. Helicopters were the thread that tied it together, flying troops into valleys, lifting casualties out of ambushes, and carrying ammunition where trucks could not go.

For the crews who had lived through Vietnam, Grenada, Panama, and Desert Storm, Afghanistan felt both familiar and new. It was another war where helicopters would decide who would survive.

Early Air Assaults and Medevac Missions

Afghanistan was a country divided by its own terrain. Valleys ran like veins through mountains higher than anything in the Rockies. Roads were few and dangerous. Snow-choked passes in winter, floods washed them away in spring, and Taliban fighters mined them in summer. Helicopters became the only way to move quickly, the only way to deliver supplies, and the only way to pull the wounded back to life.

The Chinook

At the center of these missions was the **CH-47 Chinook**. With its twin rotors, it carried more weight than any other helicopter in the American fleet. It could lift thirty or more soldiers at once, or sling artillery, Humvees, and fuel bladders beneath it. In Afghanistan's thin mountain air, it was often the only aircraft powerful enough to climb into high-altitude valleys.

Crews called it ugly but dependable. One pilot joked, "It looks like a bus with blades, but it'll carry you anywhere if you treat it right." Soldiers trusted it because it always seemed to come back, no matter the conditions.

In December 2001, helicopters carried American and Afghan fighters into the White Mountains near the Pakistan border.

The mission was to corner Osama bin Laden at **Tora Bora**, a labyrinth of caves and ridges.

The flights were long, often more than 100 miles from the nearest safe base. Chinooks went in at night, hugging the terrain, with crews wearing night-vision goggles. The altitudes pushed the engines to their limits. **CW4 Greg Calvert**, a Night Stalker pilot, recalled, "You had the power levers all the way forward and still felt like you were crawling. Every second, you thought you might clip a ridge."

Once the teams were dropped, the helicopters became lifelines. They brought water and ammunition into mountain saddles too steep for mules, and they hauled out the wounded wrapped in blankets against the cold. A Special Forces medic remembered carrying Afghan allies with burns from mortar fire into a Chinook. "We didn't have morphine left. The crew gave them water and wrapped them in jackets till we got down. Without that bird, they were gone."

Bin Laden escaped, slipping across the border, but the fight proved one thing: in Afghanistan, no mission could be done without helicopters.

Operation Anaconda

The following spring brought one of the first major set-piece battles of the war. In **March 2002**, American forces launched **Operation Anaconda** in the **Shah-i-Kot Valley**, aiming to trap hundreds of al-Qaeda and Taliban fighters.

The assault began with Chinooks carrying hundreds of infantry into landing zones across the valley. As the first aircraft descended, enemy fire erupted from the ridges. Rocket-propelled grenades and machine guns cut into the sky.

Razor 03, a Chinook carrying troops from the 101st, took a direct hit from an RPG. It crashed into the valley floor, riddled with holes. Crew chiefs pulled men out through smoke and fire. Another helicopter, **Razor 04**, was hit soon after and crash-landed. Seven Americans were killed in those first hours. Survivors remembered the pilots fighting to hold the controls steady until the last moment.

Sgt. First Class Brad Reider, who lived through one of the crashes, said, "The pilots died on the controls. They gave us seconds to get clear. Without that, none of us would walk away."

Apaches and Medevac

Above the valley, **Apache gunships** hammered enemy positions. They flew so low their rotors clipped branches. Crews returned with holes in their fuselages, some birds with half their systems shot away, and still went back.

Black Hawk medevac crews came in behind them. **Capt. Aaron Jackson**, piloting a Dustoff bird, landed under machine gun fire to pull the wounded out: his crew chief, **Spc. Daniel Ortiz** leaned out of the door and hauled soldiers aboard by their vests while rounds snapped past.

The Long Haul, 2003–2010

By 2003, the war in Afghanistan was no longer a fast campaign. Outposts dotted the valleys. Patrols ran every day. Ambushes and IEDs were constant. Soldiers on the ground came to believe in one sound above all others: the thump of helicopter blades.

Dustoff, Every Day

At Bagram, Kandahar, Jalalabad, and tiny forward bases carved into hillsides, medevac crews slept in boots and flight suits. They lived by the radio. When the call came, they ran.

Sgt. Matt Simmons, a flight medic, remembered one mission in Kunar. "He had a hole through his chest. I stayed with him the whole flight. Both hands are pressing, putting pressure to stop the bleeding. I shouted into his ear that he was not dying today. He lived." Sometimes it wasn't just modern medicine but hope, that saved lives.

Pilots dropped Black Hawks into soccer fields, dry riverbeds, and goat pastures that barely fit the rotors. Crew chiefs leaned half out of the doors, waving soldiers in, hauling litters, and calling out obstacles. They risked it because everyone on the ground believed they would.

The Mountains

The Hindu Kush tested every machine. Thin air meant less lift. To climb into some valleys, crews stripped out armor and gear to make weight.

CW3 John Keller, a Chinook pilot, said, "You pulled all the power the ship had and still felt like you were crawling. The frame shook. The engines screamed. But men were waving at you from the ridge. You held it steady and landed."

Chinooks carried thirty or more troops, or a sling load of fuel or artillery. In the thin mountain air, they were often the only birds that could do the job. Soldiers said they looked like flying buses, big and ungainly, but they trusted them. "If it was a Chinook," one grunt said, "we knew it would get there."

Dust storms were another enemy. A landing zone could vanish in seconds, nothing but a wall of brown grit. Pilots admitted they were blind. Crew chiefs stuck their heads out, shouting and using hand signals until the skids or wheels touched dirt.

The Golden Hour

By 2005, commanders pushed a new standard: the **Golden Hour**. Any wounded soldier should have surgery within sixty minutes. Helicopters made the promise real.

Lt. Jason Murray, an infantry officer, told his platoon, "If you get hit, hang on. You'll hear the birds. That sound is your ticket out."

Troops started saying they prayed for Dustoff more than for artillery. It wasn't a joke. It was a belief.

Capt. Laura Martinez, a medevac pilot, said, "We lived with helmets beside the cot. When the radio cracked, you didn't think. You just went."

The Cost of Crews

The missions never stopped. Hours piled into the hundreds. Mechanics swapped parts under floodlights, wiping sand out of engines. Crews carried the wounded and sometimes the dead. They took the weight of it too.

One crew chief remembered, "After every flight, you wiped blood off the floor. Some days it felt like you were washing the war away with rags."

Some helicopters never made it back. Taliban gunners learned to watch landing zones. Several Dustoff Black Hawks were shot down. Crews died trying to keep the promise of the Golden Hour. Each loss hit the community hard, but no one refused to take on the next mission.

Years That Would Not End

By 2010, the war had lasted longer than World War II, the Korean War, and Desert Storm. Before it was over many of the soldiers flying missions were almost as young as the conflict itself.

For Americans at home, Afghanistan was far away, something seen in brief clips on the news. For the crews, it was not distant at all. It was long nights, rotors beating through thin air, sweat freezing in mountain passes, and the sound of radios calling for Dustoff again and again.

The Later Years, 2010–2021 and The Surge

When President Barack Obama sent more than 30,000 extra troops in 2010, the tempo doubled. Helmand Province saw Marines flown in on CH-53 Sea Stallions and CH-46 Sea Knights, carried deep into valleys where roads were mined or didn't exist at all. Army Chinooks hauled fuel and artillery almost nonstop.

A Marine rifleman, **Cpl. Anthony Ruiz** said later, "The birds were part of the platoon. They dropped us off, brought us ammo, and came back for the wounded. Without them, we would have been stranded." The Chinooks made impossible missions, possible.

Crews flew until their eyes burned. Mechanics worked under floodlights, pulling sand out of filters, swapping blades, and patching holes. One crew chief remembered, "We never caught up. The war was always waiting for the next sortie."

Daily Dustoff

Medevac birds never rested. The Golden Hour standard, one hour to get a wounded soldier to a surgeon, became a creed. Black Hawks sat on strip alert around the clock.

Capt. Sarah Holden, a medevac pilot, said, "We didn't plan days. We just waited for the radio. You sleep in your boots, you eat with the headset still on your neck. When the call came, you ran."

One mission near Kandahar still haunted her. "We landed in a field wired with IEDs. Everyone knew it. But the guys were bleeding out. We took three aboard. One had both legs gone. The medic kept pressure on him until we reached the pad. He lived. That's why we go."

Soldiers on the ground knew it too. **Lt. Jason Murray**, an infantry platoon leader, said, "I told my guys, don't quit if you're hit. Keep breathing, keep fighting. You'll hear the blades. That's your ticket out."

The years dragged on. Crews rotated in and out, some serving four, five, or even six tours. Pilots said the enemy wasn't always the hardest part; it was the repetition.

CW3 Allen Briggs, a Chinook pilot, remembered, "Every mission felt the same: dust storm, high ridge, men waving at you to come down. Then you lift out with the wounded. You do it again the next night. The war just never stopped."

For medics and crew chiefs, the weight was blood. One medic said, "After a while, you don't remember faces. You remember hands holding onto yours, or boots sticking off a litter. You keep them alive till the surgeons take over."

The Drawdown

By 2014, NATO had announced that combat operations were coming to an end. On paper, it was true. On the flight lines, nothing changed. Helicopters kept moving Afghan troops, hauling supplies, and carrying Americans out as bases closed one by one.

As the drawdown spread, the missions grew riskier. Crews flew farther into valleys with less support. One pilot said, "We knew if we went down, help was hours away. But the Afghans needed us, so we went."

The Final Flights

In August 2021, the war ended the same way it had been fought, with helicopters. As Kabul collapsed, Chinooks and Black Hawks shuttled Americans and Afghan allies to the airport. The images of twin rotors beating over the embassy echoed Saigon nearly fifty years earlier.

A Marine sergeant who rode out on one of those flights said, "The cabin was shoulder to shoulder, people crying, some silent. I looked out and saw the city receding into the distance. We all knew this was the end."

The very last Dustoff flights came in the chaos of the Abbey Gate bombing. Crews landed in crowds, pulled Marines and civilians aboard, and lifted out through smoke and fire.

Memory of the Longest War

For most Americans, Afghanistan became background noise. For helicopter crews, it never faded. It was the endless grind of sorties, rotors beating through thin air, sweat and cold mixing in the same night, and the weight of carrying wounded again and again.

A flight medic put it: "We didn't win. We just kept flying. We kept pulling people out. The blades never stopped until the very last day."

Chapter Notes – Chapter 21: Afghanistan, 2001–2021
Primary Sources:

Department of Defense, *Operation Enduring Freedom After Action Reports*, 2001–2010.

U.S. Army Center of Military History, *Afghanistan: The First Phase, 2001–2002* and *Operation Anaconda: An Airpower Perspective*.

NATO/ISAF public summaries, 2003–2014.

Official Histories:

Richard Kugler, *Operation Anaconda: Lessons for Joint Operations* (NDU Press, 2007).

U.S. Army Aviation Center, Fort Novosel, oral histories of Dustoff units, 2002–2010.

U.S. Marine Corps History Division, *Marines in Afghanistan, 2001–2010*.

Published Accounts:

Sean Naylor, *Not a Good Day to Die: The Untold Story of Operation Anaconda* (2005).

Gary Schroen, *First In: An Insider's Account of How the CIA Spearheaded the War on Terror in Afghanistan* (2005).

Jake Tapper, *The Outpost* (2012), for detail on helicopter support to remote U.S. bases.

Veteran Testimonies:

Flight medic accounts collected by the Dustoff Association, including Sgt. Matt Simmons and Capt. Sarah Holden.

Ranger and Special Forces recollections of Tora Bora and Anaconda, published in Ranger Association and Special Operations journals.

Marine interviews from Helmand deployments, recorded in *Leatherneck Magazine*.

Clarification:

The Golden Hour policy was formalized in Afghanistan in 2005, driving medevac operations throughout the war.

While combat operations formally ended for NATO in 2014, U.S. helicopter crews flew missions up to the final evacuation from Kabul in August 2021.

Chapter 22 The Road To Baghdad

P relude & Context

In early 2003, Saddam Hussein stood before his generals in Baghdad and told them that Iraq would not fall. He promised another "Mother of All Battles," invoking the words he had shouted in 1991 before Desert Storm. He told his officers that American soldiers would find their graves in the desert. On state television, Iraqi propaganda showed parades of tanks and armored vehicles, soldiers goose-stepping past reviewing stands, officers raising fists to the cameras.

The reality was more complex. Saddam's army was weaker than it had been a decade earlier, worn down by sanctions and no-fly zones, but it was still formidable. He had more than **20 army divisions** and the **Republican Guard**, his elite force, stationed around Baghdad and in the south. His stockpile in-

cluded more than **2,000 tanks**, thousands of armored personnel carriers, and heavy artillery. The defenses around the capital
were layered, with armored brigades dug into revetments and
trenches. His air defense network, while degraded, still included
radar-guided surface-to-air missiles, SA-2s, SA-3s, and mobile
AAA guns.

More unpredictable was the **Fedayeen Saddam**, a paramilitary force of perhaps 20,000 men. They wore black uniforms and balaclavas, carried AK-47s, RPGs, and truck-mount-
ed heavy machine guns. They swore loyalty directly to Saddam
and promised to fight without restraint. They had orders to
ambush convoys, use civilians as shields, and turn cities into
death traps. For helicopter crews, this meant every landing zone,
every convoy escort, and every medevac run would be contested.

The Coalition Builds

In Kuwait, the desert is filled with American troops. By
March 2003, more than **250,000 Americans** were in the theater, joined by **45,000 British troops** and smaller contingents
from Australia, France, Poland, Spain, and other countries. The
U.S. Army's **3rd Infantry Division**, the armored hammer of
the invasion, staged south of the border. The **101st Airborne
Division (Air Assault)** and the **82nd Airborne** prepared
for helicopter and airborne missions. The Marines of **the 1st
Marine Expeditionary Force**, numbering more than 60,000,
staged east of Kuwait City.

On the flight lines, helicopters stretched as far as the eye
could see. Black Hawks in long rows, Apaches nose to nose,

Chinooks hulking in revetments, Marines prepping CH-53E Super Stallions with cargo nets and chains. Maintenance crews lived under their machines, pulling sand from filters, swapping out engines, balancing blades.

CW3 Mark Smith, an Apache pilot from the 101st, recalled the sense of tension. "Every night we rehearsed. Tank kills, rocket runs, low-level flying in blackout. You knew when we crossed the berm, there would be missiles, guns, and God knows what else waiting. We expected to lose birds. The question was how many."

Commanders in the Desert

The invasion would be led by **General Tommy Franks**, commander of U.S. Central Command. Franks was not a flamboyant man, unlike Norman Schwarzkopf, who had been in 1991. He was blunt, methodical, and convinced that speed and shock would win. His plan called for a lightning advance straight to Baghdad. Tanks and infantry would smash through the south while Apaches and other helicopters struck deep at Republican Guard armor.

In Britain, **Lt. Gen. John Reith** prepared his forces to secure Basra, Iraq's second-largest city. Marines would carry much of that fight. For aviators, the British contribution meant Lynx helicopters flying alongside American Cobras and Apaches.

Iraqi Confidence

Inside Baghdad, Saddam's propaganda machine broadcast daily. Anchors promised that American helicopters would fall "like rain" under Iraqi missiles. State radio claimed that every village would resist, and every street would bleed if invaded.

An Iraqi air defense officer, interviewed after the war, said, "We thought the Apaches would come like last time. We had the guns waiting. We had men with RPGs on every rooftop. We believed we could blind the Americans if we could hit their helicopters."

Life in the Camps

In Kuwait, American soldiers waited in dust and wind. Tent cities sprawled for miles. Convoys of supply trucks rolled in and out. Mechanics cursed sand that chewed through bearings and filters. Pilots cleaned their night vision goggles, crew chiefs stowed crates of ammunition in helicopters, and medics prepped litters and IV kits.

Sgt. Chris Johnson, a Chinook crew chief, remembered, "It was dust, sweat, and waiting. We slept in our flight gear. Every night, we heard Apaches practicing just over the berm. Everyone knew it was coming, and no one knew what it would look like when it started."

For infantry, the wait was just as heavy. **Pfc. Robert O'Neill** of the 3rd Infantry Division later said, "You cleaned your rifle, you played cards, you stared at the horizon. We all knew the first step across would be into history, one way or the other."

March 19, 2003

·On the night of March 19, the air war began. Cruise missiles streaked into Baghdad. Stealth fighters struck command posts. CNN broadcast live images of the capital under bombardment, anti-aircraft fire filling the sky. It was "shock and awe" replayed for the world, a storm of precision strikes meant to paralyze Saddam's command.

Within hours, ground units rolled across the berm. Tanks of the 3rd Infantry thundered north. Marines surged toward Basra. And above them, helicopters lifted in waves. Apaches skimmed the desert floor, Black Hawks carried infantry, Chinooks hauled artillery and fuel. Marines flew in on Super Stallions, their rotors beating the dawn.

The road to Baghdad had opened. For helicopter crews, it would be a gauntlet of tanks, missiles, paramilitary ambushes, and fire waiting in every city.

The Opening Air Assault

The invasion of Iraq began in the early hours of **March 20, 2003**, when tanks of the 3rd Infantry Division rolled across the berm into southern Iraq. But before the armor had gone far, helicopters were already in the fight. Apaches struck deep, Black Hawks carried infantry, and Chinooks hauled artillery and fuel.

Apaches in the Lead

The Apaches of the 11th Attack Helicopter Regiment drew some of the first missions. Their targets were the armored brigades of the Republican Guard dug in near Karbala and along the approaches to Baghdad. More than thirty AH-64s swept forward in the dark, rotors low to the sand, cockpits glowing green under night vision.

CW3 David Williams led one of the flights. "We knew the Medina Division was out there with tanks and guns," he said later. "You didn't think about the odds. You just flew low, trusted your wingman, and kept moving."

As they approached, the Iraqis opened fire. Tracers climbed out of villages, streaks of red and green across the horizon. RPGs arced into the night. Heavy machine guns rattled from rooftops.

CW4 William Stafford's Apache took a direct hit, crashed into the desert, and the crew was captured by militia. Another bird staggered back across the lines, holes stitched through its fuselage, its pilot wounded but alive.

The Apaches pressed on. Hellfires and rockets tore into the dug-in tanks. Infantry advancing the following day found burned-out hulks still smoking, the sand scorched black around them.

Sgt. David Brooks, a tank commander with the 3rd Infantry, said, "We rolled through fields of Iraqi armor already destroyed. The Apaches cleared the way. It was eerie, like walking through a graveyard of tanks."

Black Hawks Bringing In Soldiers

Behind the Apaches came **Black Hawks**, each heavy with a squad of infantrymen crouched on the cabin floor. The aircraft flew low, weaving through the fire, its door gunners leaning out with M240s.

Sgt. James Miller of the 101st remembered gripping his rifle during the flight. "You could hear the gunners firing, feel the whole ship lurch when rounds cracked past. Then we could feel the helo touch the ground, and we were on Iraqi soil. Just like that, no buildup, you were in it."

One Black Hawk, **piloted by Capt. Jason Riley** came back to base with more than thirty holes in his skin: his crew chief, **Spc. Aaron Long** said, "We patched her with tape, reloaded, and went again. Nobody sat out because of holes."

Chinooks and the Heavy Lift

The **CH-47 Chinook** was slower and bigger, but it carried what the invasion needed most: artillery, fuel, and ammunition.

CW2 Michael Torres flew his Chinook with a 155mm howitzer slung beneath. "It felt like hanging a neon sign saying shoot me," he joked later. "But those guns had to be up north before the armor moved. We were fat, slow, and noisy, but we got them there."

Another pilot, **CW3 Allen Price**, carried a load of fuel bladders beneath his ship. He recalled the fear of one hit setting the whole thing off. "You don't forget the smell of fuel and cordite mixing in the cabin. But the tanks needed gas, so you took it in."

Infantry who watched the Chinooks lumber overhead remembered the sight. **Lt. Robert Hines**, a platoon leader with the 3rd Infantry, said, "You saw those big birds carrying everything we needed. If they hadn't made it, we wouldn't have gone anywhere."

First Medevac Missions

The first casualties came within hours. RPG fire and small arms cut into convoys. Black Hawk medevac crews darted in under covering fire, crew chiefs pulling wounded aboard.

Capt. Laura Hines, a medevac pilot, recalled, "Our first pickup was a Marine near Safwan, shrapnel through his chest. We landed in a field lit by the flames of burning oil. We were down less than a minute, pulled him in, and were back in the air. He lived."

A Marine corporal who watched the pickup said later, "We thought Dustoff was a Vietnam thing. But those birds came for us, no matter what. That sound of rotors was the best thing I ever heard."

The Momentum Builds

By the end of the first week, helicopters had moved thousands of soldiers, destroyed armor, and carried dozens of wounded out of the fight. They came back with holes patched by mechanics overnight, rearmed, refueled, and launched again.

A Ranger who rode into Iraq on a Black Hawk said years later, "That flight changed me. The bird dropped us off in the dark, gunners firing, rounds hitting the skin. Then it lifted and was gone, and you realized it was just you in Iraq. No turning back."

The invasion of Iraq slowed in the southern cities. Saddam's regular army melted away, but the Fedayeen Saddam and Republican Guard dug in and fought with a ferocity that surprised many. For American troops moving north, every street corner could be an ambush, and every flight for helicopter crews meant flying through fire.

Nasiriyah

The city of Nasiriyah, straddling the Euphrates, became one of the first major fights. Marines of the **1st Marine Division** and Army supply convoys pushing through the city ran headlong into Fedayeen fighters who used mosques, schools, and homes as firing points.

A Marine platoon leader, **Lt. Shane Childers,** a graduate of The Citadel, led his men into the city on March 23. His unit came under heavy fire almost immediately. He became the first American officer killed in the war. Marines pressed forward, street by street, often fighting from rooftops and alleys.

Helicopters were called again and again. CH-46 Sea Knights and CH-53E Super Stallions lifted Marines around choke points. Black Hawks darted in for casualty evacuation.

Capt. Daniel Murphy, a Black Hawk pilot, remembered, "We were told Marines down, RPG fire everywhere. We came in low, rounds snapping through the skin. My crew chief pulled men in by their gear. We were out in less than a minute, but it felt like hours."

One Marine corporal, evacuated on that flight, said later, "I thought I was done. I couldn't breathe, couldn't move. Then I felt hands grab me, drag me in. The blades lifted, and I knew I had another chance."

The fight for Nasiriyah cost the lives of Marines and soldiers alike. By the time the city was secured, helicopters had evacuated dozens of wounded and delivered thousands of pounds of ammunition. Without the birds, many veterans thought, the advance would have stalled on the riverbanks.

The Karbala Gap

To the north, the **3rd Infantry Division** pushed toward Baghdad through the **Karbala Gap**, a narrow corridor defended by the Republican Guard's **Medina Division**. It was one of the most dangerous stretches of the invasion. Tanks and artillery were dug in, protected by trenches and revetments.

Apaches went in first. **CW2 John Hartman**, an Apache pilot, recalled, "We flew into a wall of tracers. You could feel the bird shaking with every hit. But those tanks had to go, so we pressed on."

The raid destroyed dozens of armored vehicles. Infantry who followed described a battlefield littered with burning hulks. **Lt. Paul Keller**, leading a Bradley platoon, said, "The Apaches cleared a path. Without them, we'd have lost a lot of men."

One Apache, flown by **CW4 David Alford**, was riddled with holes but limped back across the lines. His crew chief counted more than thirty impacts in the fuselage. "She was bleeding hydraulic fluid, the radios fried, but she still flew," Alford said.

Chinooks followed, carrying fuel bladders and artillery pieces. Crews joked about being flying targets, but they knew the tanks couldn't move without gas and guns.

CW3 Allen Price, a Chinook pilot, remembered, "You smelled fuel in the cabin, you knew one tracer could light it off, but the mission mattered more than the fear."

The Karbala Gap opened the road to Baghdad, and helicopter crews carried the cost and the pride of knowing they had cracked one of Saddam's strongest lines.

The Fedayeen Ambushes

The Fedayeen Saddam became the most persistent threat in the south. Dressed in black, armed with AKs and RPGs, they melted into crowds and struck from alleys. Convoys moving north were harassed daily.

Sgt. Ron Adams, a Chinook crew chief, recalled flying over a pinned convoy. "They were taking fire from both sides of the road. We dropped pallets of ammo out the back, kicking them down, and the guys on the ground waved like we'd given them their lives back."

Medevac flights into ambush zones were among the most dangerous. **Capt. Emily Carter**, piloting a Black Hawk, landed under RPG fire to pull out Marines near Nasiriyah: her crew chief, **Spc. Brian Ellis** pulled in three wounded, one missing both legs. Ellis kept pressure on the wounds the whole flight back. The Marine survived. Carter later received the **Distinguished Flying Cross**.

Another Dustoff mission near Najaf nearly ended in disaster when an RPG hit the tail of a medevac Black Hawk. The bird staggered into the air, smoking, but the pilot coaxed it back to base. The wounded made it to surgery.

Troops remembered the sound of those helicopters above them. **Cpl. James Ortega**, an infantryman with the 3rd Infantry, said, "You're in a firefight, low on ammo, pinned down. Then you hear the rotors. You don't forget that sound. It meant you had a chance to get out alive."

Keeping the Advance Moving

The battles in the south were brutal, but helicopters kept the momentum alive. They lifted artillery forward, dropped ammunition and fuel where trucks couldn't reach, and pulled out the wounded under fire.

Maj. Robert Kelly, a Marine CH-53 pilot, summed it up: "We were the bridge. Convoys stalled, roads blown, ambushes everywhere. But the birds kept moving, and because of that, the whole advance never stopped."

THE PUSH NORTH

By late March, the invasion pressed toward Baghdad. American units knew the toughest fight was still ahead: the Republican Guard divisions dug in south of the capital. Saddam had placed his best units, the **Medina**, **Hammurabi**, and **Nebuchadnezzar Divisions,** in belts of trenches, bunkers, and revetments. Iraqi propaganda promised they would "defend Baghdad to the death."

The Karbala Fight

The 3rd Infantry Division approached the **Karbala Gap**, a narrow stretch of desert between the Euphrates and Lake Razazah. Tanks, artillery, and thousands of Iraqi soldiers defended it.

On the night of March 23, more than thirty Apaches lifted off to spearhead a raid. **CW3 Greg Abbott**, flying near the front, described what they saw: "The whole horizon lit up. Tracers, RPGs, muzzle flashes out of every house. It was like the city was alive with fire."

Apaches pressed on, their Hellfires streaking across the desert. Dozens of Iraqi tanks went up in flames. But nearly every bird came back hit. One, flown by **CW4 David Alford**, limped home with hydraulics gone and bullet holes through both cockpits. His crew counted more than thirty impacts. "She should have gone down," Alford said. "But she flew us back."

Another Apache was shot down near Karbala. Its crew survived, was captured by villagers, and was later rescued. The raid was costly, but it broke the back of the Medina Division's armor.

Tankers on the ground advanced the following morning through smoking wrecks. **Lt. Paul Keller**, leading a Bradley platoon, said, "The Apaches shredded them. We rolled past burned tanks still cooking off. Without those birds, we would have paid for every yard."

Black Hawks and Chinooks

While Apaches killed tanks, Black Hawks and Chinooks kept the advance moving. Black Hawks lifted infantry into blocking positions. Chinooks carried artillery and fuel forward.

CW2 Daniel Reeves, a Chinook pilot, remembered one run with a 155mm howitzer slung beneath. "We were fat and slow, and everyone was shooting at us. But that cannon had to be there before daylight. We dropped it and got out, and the next morning it was firing over the guys' heads as they pushed."

Crew chiefs leaned out with their machine guns, firing at muzzle flashes below. Infantry remembered the sight. **Sgt. Anthony Bell**, a Bradley gunner, said, "Every time we thought we were out of ammo, the Chinooks came overhead with more. That kept us in the fight."

Dustoff Missions

The push north bled men. Fedayeen ambushes cut convoys, artillery found its mark, and RPGs hit vehicles. Medevac crews flew straight into the chaos.

Capt. Mark Hall, a Dustoff pilot, landed in a burning field near Najaf to pull soldiers hit by mortars. "I couldn't see ten feet. My crew chief talked me down, hand signals in the smoke. We loaded fast, and we were off."

Sgt. Lisa Green, the medic, sat on the floor working over a Marine with shrapnel in his chest. "He squeezed my hand once and passed out. I told him he'd wake up in a hospital, and he did."

Another crew, flying into the Karbala Gap, picked up a squad leader with both legs shattered. The medic cut his pants off with a knife, slapped on tourniquets, and kept talking to him over the roar of the rotors. He lived.

Breaking the Guard

By the first days of April, the Republican Guard was broken. Apaches had burned their tanks, Abrams and Bradleys cut through their trenches, and helicopters had carried supplies and wounded nonstop. Survivors of the Guard later admitted they had never faced anything like the Apaches. One Iraqi officer said after capture, "Your helicopters came like demons. We fired everything, and they still killed our tanks."

A64 Apache in Iraq Public Domain

Infantry pressing north saw the evidence. **Cpl. Robert Hines** said, "We rolled past fields of wrecked armor, and overhead the Apaches were still circling. We knew then Baghdad was open."

The Human Side

For crews, the push north meant exhaustion. Pilots flew twenty hours at a time. Crew chiefs counted holes in the fuselages and taped them over before dawn. Medics scrubbed blood from the decks after every run.

One Black Hawk crew chief, **Spc. Aaron Long** said, "You landed, patched holes, loaded more ammo, and flew again. There wasn't time to think. You just kept the blades turning."

For the soldiers on the ground, it was a reassuring sign. **Pfc. Daniel Ortiz**, 3rd Infantry Division, put it best: "Every time you thought you were cut off, the birds showed up. They carried ammo, fuel, and the wounded; it didn't matter. You heard them and you knew you weren't alone."

The Thunder Run and Baghdad's Fall

By early April 2003, the gates of Baghdad had been opened. The Republican Guard was broken, its armor burning in the desert south of the capital. What remained were city streets, ambushes, and the final gamble of driving armored columns into the heart of Saddam Hussein's city.

The First Thunder Run

On **April 5**, a column of Abrams tanks and Bradleys from the 3rd Infantry Division roared up Highway 8 and into Baghdad. This was not a probing patrol. It was a charge straight into the capital.

RPGs streaked from overpasses. Machine guns rattled from alleyways. Technicals, pickup trucks mounted with heavy weapons and manned by Saddam's militia, darted out of side streets. But the tanks pressed on.

Overhead, Apaches fired down alleys, their 30mm cannons shattering bunkers. Black Hawks dropped infantry into cloverleafs and intersections to block counterattacks. Chinooks carried fuel forward, the big birds flying slow and vulnerable but essential.

Sgt. First Class John Wolff, a Bradley commander, said later, "It was like nothing I'd ever seen. Tanks firing in every direction, Apaches overhead, the whole city alive with fire. But we didn't stop. We kept going."

By nightfall, the armor pulled back, proving that the heart of Baghdad was within reach.

The April 7 Thunder Run

Two days later came the decisive strike. On **April 7**, the 2nd Brigade of the 3rd Infantry pushed all the way to the presidential palaces along the Tigris River.

The column smashed through ambushes at every turn. RPGs struck Bradleys, artillery fell blind across highways, and Saddam's Fedayeen fought from alleys. But the armor did not stop. Tanks rolled past the Ministry of Defense, through Republican Guard headquarters, and into Saddam's opulent palace grounds.

Col. David Perkins, the brigade commander, said, "Every block was a fight, but the momentum was ours. By afternoon, we were parked in Saddam's front yard."

From the air, Baghdad looked like a city in flames. **Sgt. Lisa Green**, a Black Hawk medic, remembered, "Columns of smoke everywhere, tracer fire arching into the sky, and our birds dropping into intersections with rounds snapping past. Every time we landed, soldiers waved us in because they knew Dustoff meant survival."

The Collapse of a Regime

Even as American tanks occupied Baghdad, Saddam's propaganda minister, **Muhammad Saeed al-Sahhaf**, stood before cameras declaring that coalition forces were being destroyed. Dubbed "Baghdad Bob" by U.S. troops, he claimed American soldiers were committing suicide at the gates even as Abrams tanks thundered behind him. Soldiers listening to radios laughed in disbelief. One officer said, "We were rolling through his city while he was on TV saying we didn't exist."

The regime was cracking. Soldiers abandoned their uniforms. Officers fled. Fedayeen fighters melted into neighborhoods. Baghdad was falling apart even as the armored columns held its heart.

The Crossed Swords

The victory arch, a remnant of the Iraq-Iran War, symbolized the entrance to Baghdad. It was a huge edifice of crossed swords, made from melted-down Iranian weapons, captured during the conflict with the Ayatollah's forces.

Now American tanks rumbled beneath it. Helicopters circled above, watching the same streets where Iraqi parades had once marched in triumph. For U.S. soldiers, it was surreal. **Pfc. Daniel Ortiz** said, "You grew up seeing pictures of that arch in Time magazine. Then suddenly you're rolling under it with your rifle in your lap, and you realize Saddam's time is over.

The Statue Falls

One of the symbols of Saddam's reign was his statue in Fidro Square. As American troops approached, an Iraqi crowd emerged around the statue as it suddenly became a symbol of their suffering over the past years, and they tried to bring it down. Finally, with the aid of a tank recovery vehicle, they were successful.

From the air, Black Hawk crews saw it live. **CW2 Aaron Kelley**, circling above, said, "The whole square erupted. You could see people cheering, climbing on the pieces, and slapping the head of Saddam with shoes, an Iraqi insult. It was history happening right under us."

Television crews caught the statue falling live, and with it the mass of Iraqis celebrating the regime's fall, surrounded by American soldiers with their weapons at sling arms.

For the infantry who fought their way into the city, the memory was one of chaos and disbelief. **Spc. Eric Compton**, a 3rd Infantry rifleman, said, "You expected block-by-block house fighting. Instead, it was RPGs, looters, fire everywhere, and then suddenly we were sitting in Saddam's palace gardens eating MREs."

A Marine corporal recalled, "You walked into rooms with gold ceilings and chandeliers, and outside you could still hear gunfire. It felt like two worlds smashed together."

Baghdad Falls

By the evening of April 9, the regime had collapsed. Saddam and his sons had fled. Ministries burned. Looters stormed government buildings. American forces held the city. Helicopters still flew, carrying supplies and wounded, circling over a capital that was no longer in Saddam's hands.

For pilots and soldiers alike, it was an unforgettable experience. One Apache pilot summed it up: "We fought through fire, tanks, ambushes, and by the end, we were flying over Baghdad watching the statue come down. You knew then the war had changed the world."

Chapter Notes Iraq: The Road to Baghdad, 2003
Primary Sources:

U.S. Army Center of Military History, *On Point: The United States Army in Operation Iraqi Freedom* (2004).

Department of Defense, After Action Reports, 3rd Infantry Division and 1st Marine Expeditionary Force, March–April 2003.

U.S. Army Aviation Digest, coverage of Apache, Black Hawk, and Chinook operations during OIF I.

Official Histories:

U.S. Marine Corps History Division, *U.S. Marines in Iraq, 2003: Anthology and Annotated Bibliography.*

Joint Center for Lessons Learned (JCLL), *Operation Iraqi Freedom Lessons Learned* (2003–2004).

U.S. Air Force Historical Studies, *Shock and Awe and the Opening of OIF.*

Published Accounts:

Rick Atkinson, *In the Company of Soldiers: A Chronicle of Combat* (2004).

Thomas Ricks, *Fiasco: The American Military Adventure in Iraq* (2006).

David Zucchino, *Thunder Run: The Armored Strike to Capture Baghdad* (2004).

Veteran Testimonies:

Oral histories of Apache pilots from the 11th Aviation Regiment, including CW3 Greg Abbott and CW4 David Alford, collected by Army Aviation Association of America (AAAA).

Ranger, infantry, and armor testimonies of Thunder Run participants, published in Ranger Association journals and unit reunion records.

Dustoff Association accounts from medevac crews who flew missions at Nasiriyah, Karbala, and Baghdad.

Marine CH-53E and CH-46 crews interviewed in *Leatherneck Magazine* (2003–2005).

Clarification:

Coalition forces totaled more than 300,000, with U.S. Army, Marine, and British divisions leading the drive.

Casualty figures for OIF I: U.S. losses were 172 killed and over 500 wounded in the major combat phase.

The Republican Guard Medina and Hammurabi Divisions were effectively destroyed by early April, clearing the way for the Thunder Runs.

The fall of Baghdad on April 9, 2003, was marked by the toppling of Saddam's statue in Firdos Square, broadcast worldwide as a symbol of regime collapse

CHAPTER TWENTY-THREE

IRAQ, The Long Stay

The shooting war in 2003 was fast. Baghdad fell in weeks, and Saddam's statues were pulled down in the squares. But that wasn't the end, it was the start of something more complex. What followed was the long, grueling fight against insurgency. In that kind of war, helicopters never went away. Every day, every night, the sound of rotors hung over Iraq, carrying troops, hauling supplies, pulling the wounded out of ambushes.

The First Losses

In November 2003, only months after Baghdad fell, a **CH-47 Chinook** packed with soldiers heading for leave was shot down near Fallujah. A surface-to-air missile slammed into the big helicopter, killing 16 and wounding more than 20. It was the single deadliest attack on American aviation since Vietnam.

Survivors remembered the chaos. **Sgt. Randy Robles**, who crawled from the wreckage, said, "We thought we were going home. Then there was fire everywhere, screaming, metal ripping apart. I don't know how I made it out."

The shootdown sent shock waves through every flight line in Iraq. Crews knew the enemy was adapting. Helicopters became prime targets for insurgents with RPGs and MANPADS.

Baghdad from Above

By 2004, the war had shifted into counterinsurgency. Convoys were attacked daily by roadside bombs and small arms fire. To reduce casualties, commanders leaned on helicopters. Black Hawks became the taxis of the war, shuttling troops across Baghdad rather than risk the streets.

CW3 Kevin White, a Black Hawk pilot, said, "Every time we lifted, you felt like the whole city was staring at you. You flew low and fast, because you knew there were guys down there with RPGs waiting."

Crew chiefs fired out the doors with M240s, watching alleys and rooftops. Soldiers strapped into canvas seats remembered the flights as both terrifying and comforting. **Pfc. Andrew Lewis**, riding over Baghdad in 2005, said, "You wanted the bird because you knew the roads were worse. But you never forgot you were a target in the sky."

Dustoff in the Insurgency

Medevac flights were relentless. IEDs shredded Humvees and Strykers daily. Snipers hit patrols. The Golden Hour standard still held, and crews risked everything to reach the wounded.

Capt. Jennifer Morgan, a Dustoff pilot, recalled landing in Sadr City at night to evacuate a soldier hit by a sniper. "The LZ was lit up with tracers. We came down between buildings, rounds snapping across the canopy. My crew chief hauled the casualty aboard. We lifted out low, and I remember thinking we might not make it. But we did, and the kid survived."

A medic, **Sgt. James Baker** said, "You worked on boys younger than you were. Tourniquets, morphine, chest seals, whatever you had. You promised them they'd get home. Some didn't, but many did because we got there fast."

Anbar and the Long Road

In western Iraq, the Marines fought through **Anbar Province**, Fallujah, Ramadi, and Haditha. Helicopters carried them into combat again and again. CH-46s and CH-53s lifted rifle companies into dusty landing zones, often under fire. Apaches circled overhead, pounding insurgent strongholds with rockets and cannon.

Maj. Robert Ellis, a Marine CH-53 pilot, said, "We flew into places that looked like the surface of the moon. Dust clouds, gunfire, RPG trails. You dropped your Marines, lifted, and just hoped you could come back for them."

Medevac birds followed close behind. During the **Second Battle of Fallujah in 2004**, dozens of Dustoff flights carried Marines out of the city's streets. One crew chief recalled dragging bloodied Marines aboard, the cabin shaking with cannon fire overhead. "They were kids. You held them down and told them they'd see home again. That was the only thing you could give them."

The Grind

By 2006, Iraq had become the longest sustained helicopter war since Vietnam. Crews flew hundreds of hours each month. Pilots rotated through multiple tours. Crew chiefs reported feeling the vibration of the rotors even when they lay down to sleep.

CW2 Patrick Long, after three tours flying Black Hawks, admitted, "You stopped counting missions. You just kept flying. Baghdad, Ramadi, Mosul, it all blurred together. But you kept flying because somebody was always waiting."

For soldiers on the ground, the sound of blades remained the sound of survival. One infantry sergeant said, "We joked that God wore rotor blades. You'd be bleeding in a ditch, and you'd hear them coming. That meant you had a chance."

The Later Years, 2007–2011

By 2007, the U.S. had been in Iraq for four years. Insurgents controlled neighborhoods, IEDs tore through convoys daily, and sectarian violence ripped Baghdad apart. President George W. Bush ordered a surge of more than 20,000 troops. For helicopter crews, it meant the war only grew heavier.

The Surge

The surge brought constant missions. Black Hawks lifted platoons into Baghdad at night, their rotors kicking dust into abandoned streets. Chinooks carried pallets of ammunition and stacks of bottled water to forward outposts cut off by ambushes. Apaches prowled over the city, their cannons hammering insurgent hideouts in alleys and courtyards.

CW2 Brian Keller, a Black Hawk pilot, remembered, "You felt like the city never slept. Every flight, someone was shooting at you. You went low, fast, and just prayed the RPG missed."

In Diyala Province, medevac crews flew almost nightly. Patrols walked into IED blasts or sniper fire, and Dustoff calls came in steady.

Capt. Laura "Doc" Hanson, a medevac pilot, said, "We landed in streets where the power was out, where you couldn't tell if the crowd was trying to help you or kill you. The medic dragged the wounded in, and we lifted before the next round came."

The Toll on Crews

For many crews, the strain was endless. Pilots flew hundreds of hours a month. Mechanics patched bullet holes with sheet metal and tape, pulling all-nighters under work lights. Crew chiefs lived with the sound of the rotors in their bones.

Sgt. Kevin Long, a Chinook crew chief, said, "We carried everything. Fuel, ammo, prisoners, you name it. It felt like we were the veins of the war. And we never stopped."

Medevac medics carried the heaviest weight. **Sgt. Maria Torres** recalled holding a 19-year-old infantryman who had lost both legs to an IED. "He kept asking if he was going home. I told him yes, even though I didn't know if he'd make it. He squeezed my hand all the way until we hit the pad. He survived."

Drawdown

By 2009, the surge had slowed the insurgency, but the war was not over. Bases began to close. Outposts shrank. Convoys still faced bombs on the roads, and helicopters remained the lifeline.

Black Hawks shuttled troops between shrinking bases. Chinooks carried dismantled equipment back south. Medevac flights still answered calls nightly. Soldiers joked that the last thing left in Iraq would be helicopters circling overhead.

CW4 Patrick Reeves, an Apache pilot, said, "We were the last line of support. When the bases pulled back, the birds continued to fly. It felt like we were holding the whole war together with rotor blades."

The Final Flights

In December 2011, after eight long years, the last U.S. forces left Iraq. In the final weeks, helicopters carried out missions that looked the same as they had in 2003, troops loaded with gear, pallets of supplies slung beneath Chinooks, medevac birds landing in dust to lift the wounded.

Sgt. Thomas Lane, a Black Hawk crew chief, recalled the last time he flew out of Baghdad. "We lifted in the dark, same as always, and circled the city one last time. You could see the Tigris shining. We'd lost friends here, we'd bled here, and now it was over. It felt unreal."

When the last American convoy crossed into Kuwait on December 18, 2011, helicopters still orbited overhead, covering their exit. Then they, too, turned south.

Memory of the Long Stay

For the crews, Iraq was not remembered in speeches or dates but in missions that never ended. One Dustoff medic said, "I don't remember how many we saved. I remember faces, the weight of bodies on litters, the sound of blades carrying us through the dark."

For the soldiers who rode the birds, the sound of rotors was the thread of survival. A Marine said simply, "The roads belonged to the enemy. The sky belonged to us. That's what got us home."

Chapter Notes – Iraq: The Long Stay, 2003–2011

Primary Sources:

Department of Defense, *Operation Iraqi Freedom After Action Reports* (2003–2011).

U.S. Army Center of Military History, *On Point II: Transition to the New Campaign* (2008).

1st Cavalry Division and 82nd Airborne Division rotation summaries, Baghdad and Anbar deployments.

Official Histories:

U.S. Army Medical Department, *Golden Hour in Iraq and Afghanistan* (2010).

U.S. Marine Corps History Division, *U.S. Marines in Iraq, 2004–2008.*

Joint Center for Lessons Learned (JCLL), aviation studies on CH-47 and UH-60 operations in Iraq.

Published Accounts:

Bing West, *No True Glory: A Frontline Account of the Battle for Fallujah* (2005).

Michael Gordon & Bernard Trainor, *Cobra II* (2006).

Richard Engel, *War Journal: My Five Years in Iraq* (2008).

Veteran Testimonies:

Oral histories from CH-47 crew chiefs and UH-60 medevac pilots collected by the Dustoff Association, 2003–2011.

Marine CH-53 and CH-46 crews interviewed in *Leatherneck Magazine*, especially around Fallujah   and Ramadi operations.

Infantry and medevac accounts from the Second Battle of Fallujah, compiled in unit reunion records.

Interviews with families of soldiers lost in helicopter shootdowns (Fallujah, 2003–2005).

Clarification:

The deadliest helicopter loss of the war was November 2003 near Fallujah, when a CH-47 was downed by a missile, killing 16 soldiers.

Medevac "Golden Hour" standards were enforced even at the height of the insurgency, with crews often flying into small arms and RPG fire.

The final U.S. combat troops left Iraq on December 18, 2011, with helicopters providing overwatch for the last convoys exiting into Kuwait.

Epilogue

The story of the helicopter in war began with a fragile machine of fabric and wood in the jungles of Burma. **Lt. Carter Harman** coaxed his YR-4B into the sky in 1944 and lifted out a handful of wounded airmen. It was noisy, slow, and underpowered. But it worked. For the first time, men who would have been left behind were flown out. That moment lit a fire that has never gone out.

In Korea, the H-13 Sioux became a lifeline. Pilots lashed stretchers to the skids and carried the wounded from frozen ridges and rice paddies. At the Chosin Reservoir, helicopters kept men alive when the cold and the Chinese assault threatened to kill them all. The Sioux was crude, but it proved the idea.

In Vietnam, helicopters became the heartbeat of the war. The Huey was everywhere, bringing soldiers in, pulling the wounded out, circling above with gunners firing. **Major Charles Kelly** set the standard with his words, *"When I have your wounded,"* before he was killed at the controls. His creed became the

Dustoff creed, carried by every crew that flew into fire for the rest of the war.

Vietnam was the crucible. Helicopter pilots and medics rescued thousands of men from hot zones. Many never came back themselves. More than 200 Dustoff crewmen died in Vietnam trying to keep the promise. For every man lifted, for every soldier who made it to a field hospital alive, their sacrifice was written in blood.

The story did not end there. In 1993, the Night Stalkers flew into Mogadishu. Two Black Hawks were shot down, crews killed or captured, Gordon and Shughart fighting to the last bullet to protect their brothers. Their courage, their refusal to quit, became part of the same story, the same promise Kelly had spoken three decades earlier.

In Iraq, helicopters carried the invasion north and kept the fight alive through the long insurgency. Apaches tore apart Republican Guard armor in the desert. Chinooks carried fuel and artillery, flying straight into the fire as big targets. Black Hawks shuttled troops over Baghdad, safer than the roads below. Dustoff came into alleys and ambush sites, pulling men out of burning trucks. The war stretched for years, and the rotors never stopped turning.

In Afghanistan, helicopters became the primary means of fighting. The mountains swallowed convoys. The roads were mined and deadly. It was Chinooks climbing into thin air, Apaches circling valleys, Black Hawks landing on goat tracks, medevac crews carrying boys with shattered bodies out of firefights. From Tora Bora to Anaconda, through the surge and the drawdown, helicopters were the thread tying the war together.

From Burma to Baghdad, from Chosin to Kandahar, one truth ran through them all. Helicopters were more than machines. They were lifelines. They carried courage in their cabins and left sweat and blood on their deck plates. They carried the wounded, the scared, the dying, and they brought the promise of survival.

A pilot once said, "We didn't win or lose wars. We just kept the promise. We came when they called." That was enough.

The names and places change, but the sound remains the same. The thump of blades overhead, the sudden rush of air as a helicopter comes in under fire, the hands pulling a soldier aboard. That sound is the sound of hope, the sound of survival.

When I have your wounded.

That was the promise made by Kelly in Vietnam, and kept by every crew that followed, in Mogadishu, in Iraq, in Afghanistan. It will be the promise as long as helicopters fly into battle.

This first volume tells the story of Dustoff pilots, slick crews, gunships, and SeaWolves, as well as the mercy missions of Vietnam. But there is another part of history. While helicopters worked the paddies and canals in the south, another community flew into the most dangerous skies of the war. The Air Force's Jolly Green Giants ranged deep into North Vietnam. The Navy's Big Mothers and Clementines covered carrier pilots and sailors in the Gulf of Tonkin and coastal areas of North Vietnam. The

Air Force's Pedro crews, flying HH-43 Huskies, answered short-range rescue calls from bases and battle zones day after day.

That story, of combat search and rescue over the north, of pilots and pararescuemen who braved SAM belts and MiG interceptors to bring men home, deserves its own telling.

This will be detailed in *Above and Beyond, Volume 2: Combat Search and Rescue over North Vietnam*, which will be available soon. Look for it in the spring of 2026.

From the old, as in very old, Navy helo crewman that wrote this, my apologies to my Huey pals for anything I got wrong, sorry guys. Dear Reader, the self-publishing world is driven by reviews, if you think this was worth your time please leave a review, if on the other hand you believe that reading of exploits from so many decades past was a waste of your time, tell that part also. These wonderful people in this work deserve nothing less.

Acknowledgements

I need to thank all of those wonderful guys that offered their existence and all of their tomorrows in order to save a few fellow countrymen.

This was not an outline designed narrative, instead it just existed from a seat of the pants flight. Hopefully this old Geezer (77 at time of writing) got most of it correct. If I missed some names, my apologies. There were so many flights of self-sacrificing courage, and I know I missed more that i hit. When I was younger and had hair and one chin, I was privileged to fly in the Sikorsky SH 3A and HH 3A, but the Huey was, as Joe Galloway put it, the sound of the Vietnam War, with its two bladed unique sound. Guys like Bill Cope, whom I met, while coaching in Abbott, Texas, made these aircraft the revolution of airborne warfare.

This was never an organized narrative, but more a chain of thoughts. What the helicopter, that marvelous flying machine first envisioned by daVinci so many centuries ago, has done to make the battlefield ambulance a quicker more efficient in its constant search for The Golden Hour. Major Kelly, we still hear you. "WHEN I HAVE YOUR WOUNDED"

I, as a former High School dropout, and later college graduate, need to thank the makers of Grammarly for upgrading my English skills in this endeavor. Yes, it is a function of computer-based intelligence, and I used both it and Microsoft Word for honing some of the punctation and sentence structure.

www.ingramcontent.com/pod-product-compliance
Lightning Source LLC
Chambersburg PA
CBHW060404130626
46555CB00005B/1989